| era | period | | age/stage/series | 230 |
|---|---|---|---|---|
| UPPER PALAEOZOIC | Permian | Upper | Tatarian<br>Kazanian | |
| | | | | 280 |
| | Carboniferous | | | 345 |
| | Devonian | Upper | Famennian<br>Frasnian | |
| | | Middle | Givetian<br>Couvinian | |
| | | Lower | Emsian<br>Siegenian<br>Gedinnian | 400 |
| LOWER PALAEOZOIC | Silurian | Upper | Přidolian<br>Ludlow<br>Wenlock | |
| | | Lower | Llandovery | 435 |
| | Ordovician | Upper | Ashgill<br>Caradoc | |
| | | Lower | Llandeilo<br>Llanvirn<br>Arenig<br>Tremadoc* | 500 |
| | Cambrian | Upper | | |
| | | Middle | | |
| | | Lower | | 570 |
| | Pre-Cambrian | | | |

*included with the Upper Cambrian in some classifications.



**NOTES** The age and stage names given here for the various geological systems are in common use in western Europe, but many other classifications are also used.

The meanings of the words *era*, *sub-era*, *period*, *epoch*, *age*, and *stage* will be found on page 113.

Ages for the boundaries between the geological systems are given in millions of years (Ma).

ALEC WATT

# BARNES & NOBLE THESAURUS OF GEOLOGY

the principles of geology
explained and illustrated

BARNES & NOBLE BOOKS
A DIVISION OF HARPER & ROW, PUBLISHERS
New York, Cambridge, Philadelphia, San Francisco
London, Mexico City, São Paulo, Sydney

**Library of Congress Cataloging in Publication Data**

Watt, Alec.
The Barnes & Noble thesaurus of geology.

Includes index.
1. Subject headings—Geology. I. Title.
Z695.1.G43W37 1983 025.4'955 82-48833
ISBN 0-06-015177-3
ISBN 0-06-463579-1 (pbk.)
Printed in Spain by Heraclio Fournier SA

# Contents

**How to use this book** *page* 5

**The earth** 9

**Geophysics** 11
Gravity; earthquakes & seismology; the Earth's magnetism

**Geochemistry** 15
Elements, ions, molecules; chemical compounds; reactions, classification; physical chemistry

**Weathering and erosion** 20
Processes; effects; wind erosion

**Soils** 23

**Streams and rivers** 24
Stream erosion; valleys; floodplains & deltas; drainage patterns

**Ice and ice action** 28
Glaciers, ice sheets; erosion, deposition; marginal effects, glaciated valleys

**Geomorphology and land forms** 32
The erosional cycle; surface features

**The oceans** 34
Physical features; water movements & deposits

**Coasts** 37
Physical features; development

**Crystallography** 40
Crystal forms & measurement; symmetry; crystal systems

**Minerals** 44
General properties; descriptive terms; relationships; optical properties; native elements & oxides; phosphates; sulphides; carbonates; sulphates, tungstates, halides; silicate structures; silica minerals, micas, ferromagnesian minerals; feldspars; pyroxenes, amphiboles; olivines, garnets, feldspathoids; aluminium & other silicates; hydrated silicates, clay minerals

**Igneous petrology** 62
General; intrusions; volcanoes; composition diagrams; descriptive; igneous rocks; classification; rock types

**Sediments** 80
Deposition & stratification; sedimentary environments; sedimentary structures; lithification & diagenesis

**Sedimentary rocks** 85
Lithology; carbonates & chert; arenites, rudites; argillaceous & ferruginous rocks; carbonaceous rocks & hydrocarbons

**Metamorphism** 90
General; metamorphic grade; metamorphic facies, contact effects; effects of pressure & heat; textures & structures

**Metamorphic rocks** 96

**Palaeontology** 98
General; taxonomy; palaeoecology; evolution; invertebrates; chordates; fossil plants; stratigraphical palaeontology

**Stratigraphy** 112
General; time & other divisions; geological systems; periods of mountain-building; zoning & correlation; relationships

**Geological time** 120

**Structural geology** 122
General; dip, strike, & folds; faults, thrust & nappe tectonics; miscellaneous structures; large-scale structures

**Plate tectonics** 134
General; plate margins, island arcs; constructive margins; destructive margins; palaeogeography; convection

**Engineering geology** 143

**Oil geology** 144

**Mining geology** 145

**Hydrogeology** 146

**Field work and laboratory work** 147

**Meteorites** 149

**Geology of the moon** 150

**Appendixes:**
*One:* Additional definitions 151
*Two:* Common abbreviations in geology 162
*Three:* International System of Units (SI) 163
*Four:* Understanding scientific words 165

**Acknowledgements** 170

**Index** 171

# How to use this book

This book contains some 1500 words used in the geological sciences. These are arranged in groups under the main headings listed on pp.3–4. The entries are grouped according to the meaning of the words to help the reader to obtain a broad understanding of the subject.

At the top of each page the subject is shown in bold type and the part of the subject in lighter type. For example, on pp.12 and 13:

**12 · GEOPHYSICS**/EARTHQUAKES AND SEISMOLOGY

**GEOPHYSICS**/EARTHQUAKES & SEISMOLOGY · **13**

In the definitions the words used have been limited so far as possible to about 1500 words in common use. These words are those listed in the 'defining vocabulary' in the *New Method English Dictionary* (fifth edition) by M. West and J.G. Endicott (Longman 1976). Words closely related to these words are also used: for example, *characteristic*, defined under *character* in West's *Dictionary*. For some definitions other words have been needed. Some of these are everyday words that will be familiar to British readers; others are scientific words that are not central to geology. These are listed in Appendix 1 (pp.151–61). If therefore you find a word in a definition that is not familiar to you, you should turn to Appendix 1.

## 1. To find the meaning of a word

Look for the word in the alphabetical index at the end of the book, then turn to the page number listed.

The description of the word may contain some words with arrows in brackets (parentheses) after them. This shows that the words with arrows are defined near by.

(↑) means that the related word appears above or on the facing page;

(↓) means that the related word appears below or on the facing page.

A word with a page number in brackets (parentheses) after it is defined elsewhere in the dictionary on the page indicated. Looking up the words referred to in either of these two ways may help in understanding the meaning of the word that is being defined.

The explanation of each word usually depends on knowing the meaning of a word or words above it. For example, on p.123 the meaning of *axial plane, fold-axis,* and the words that follow depends on the meaning of the word *fold*, which appears above them. Once the earlier words are understood those that follow become easier to understand.

## 2. To find related words

Look in the index for the word you are starting from and turn to the page number shown. Because this book is arranged by ideas, related words will be found in a set on that page or one near by. The illustrations will also help here.

For example, words relating to volcanic eruptions are on pp.68–70. On p.68 *volcano* is followed by words used to describe various kinds of volcanic eruption and types of volcano; p.69 give words for solid material thrown out from volcanoes, p.70 lists words for liquid and gaseous materials from volcanoes.

## 3. As an aid to studying or reviewing

There are two methods of using this book in studying or reviewing a topic. You may wish to see if you know the words used in that topic or you may wish to review your knowledge of a topic.

(*a*) To find the words used in crystallography, you would look up *crystallography* in the alphabetical index. Turning to the page indicated, p.40, you would find *crystal, crystallize, crystallography, crystal lattice*, and so on. Turning over to p.41 you would find *crystallographic axis, intercept*, and so on; and on p.42 *symmetry* etc.

(*b*) Suppose that you wished to review your knowledge of a topic; e.g. *sediments*. If, say, the only word you could remember was *deposit* you could look in the alphabetical index and find *deposit*. The page reference is to p.80. There you would find the words *sediment, sedimentation, deposit, deposition, bed, stratum, bedding-plane*, etc. If you next looked at p.81, you would find words relating to the various environments in which sediments are deposited; then on p.82 you would find words relating to the characters of sediments; on p.83 words describing sedimentary structures; and so on.

## 4. To find a word to fit a required meaning

It is almost impossible to find a word to fit a meaning in most dictionaries, but it is easy with this book. For example, if you had forgotten the word for the angle between two axis of a fold and the horizontal all you would have to do would be to look up *fold axis* in the alphabetical index and turn to the page indicated, p.123. There you would find the word *pitch* with a diagram to illustrate its meaning.

# THE THESAURUS

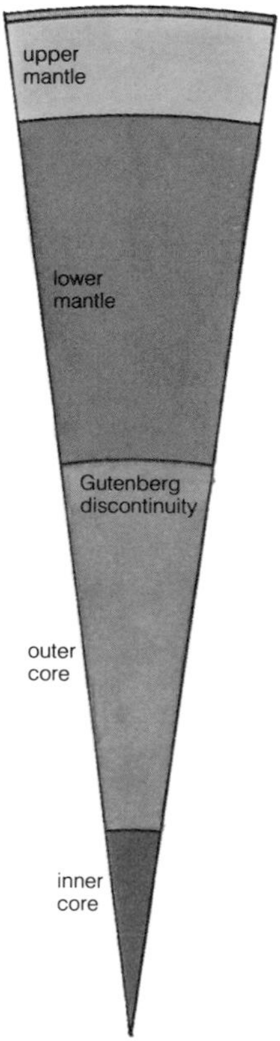

**geology** (*n*) the science of the Earth: how it was formed, what it is made of, its history and the changes that take place on it and in it. Geology includes parts of geophysics (p.11), mineralogy (p.44), petrology (p.62), stratigraphy (p.112), palaeontology (p.98), and structural geology (p.122). **geological, geologic** (*adj*).

**Earth sciences** a group of sciences that includes geology, geophysics (p.11), geochemistry (p.15), oceanography (p.34), meteorology (the study of the weather), and astronomy (the study of the heavenly bodies) so far as it concerns the Earth.

**crust** (*n*) the part of the Earth above the Mohorovičić discontinuity (p.10). It is less dense than the mantle (↓). The *continental crust* of the great land areas is thicker, less dense, and older than the *oceanic crust*. **crustal** (*adj*).

**mantle** (*n*) the part of the Earth between the crust (↑) and the core (↓), i.e. between the Moho (p.10) and the Gutenberg discontinuity (p.10). It probably consists largely of $MgO$ and $SiO_2$ with sodium, calcium, and aluminium.

**core** (*n*) the central part of the Earth, below the Gutenberg discontinuity (p.10) at a depth of about 2900 km below the Earth's surface. The core is thought to consist almost entirely of iron. It can be divided into the *outer core*, which may be liquid, and the *inner core*, which may be solid, at a depth of 5100 km. The density (p.154) of the core is more than twice the density of the mantle (↑).

**lithosphere** (*n*) the outer, solid part of the Earth: the crust (↑) and the upper part of the mantle (↑) to a depth of about 100 km. The lithosphere is stiffer than the asthenosphere (↓). **lithospheric** (*adj*).

**asthenosphere** (*n*) the part of the mantle (↑) from a depth of about 100 km to 250–300 km. It is not as strong and stiff as the lithosphere (↑).

**mesosphere** (*n*) the part of the mantle (↑) below the asthenosphere (↑), i.e. from a depth of 250–300 km to the core (↑).

**atmosphere** (*n*) the gases surrounding the Earth. **atmospheric** (*adj*).

crust
upper mantle
lower mantle
Gutenberg discontinuity
outer core
inner core

**section through the earth**

**sial** (*n*) a term for the parts of the Earth's crust (p.9) made up of rocks containing silica (p.16) and alumina (p.16). **sialic** (*adj*). *See also* **sima** (↓).

**sima** (*n*) a term for the parts of the Earth's crust (p.9) made up of rocks containing silica (p.16) and magnesium. **simatic** (*adj*). *See also* **sial** (↑).

**discontinuity** (*n*) a layer or boundary within the Earth that separates parts of the Earth having different properties, e.g. seismic properties. See also Mohorovičić discontinuity (↓) and Gutenberg discontinuity (↓).

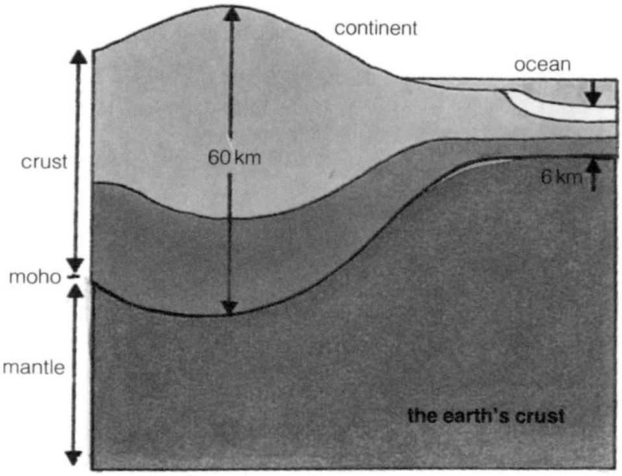

**Mohorovičić discontinuity, Moho, M discontinuity** a boundary that separates the crust (p.9) above from the mantle (p.9) below. The Moho is at a depth of about 20–40 km below the surface of the continents and about 10 km below the ocean floor. There is a difference between the velocities of earthquake waves (p.12) above and below the Moho.

**Gutenberg discontinuity** a boundary that separates the mantle (p.9) from the core (p.9) at a depth of about 2900 km below the Earth's surface. The velocities of earthquake (p.12) waves are different above and below the Gutenberg discontinuity.

**Weichert–Gutenberg discontinuity** = Gutenberg discontinuity (↑).

**geophysics** (*n*) the study of the physics (p.158) of the Earth, including the hydrosphere (p.34) and the atmosphere (p.9). **geophysical** (*adj*).

**gravity** (*n*) the force that pulls a body towards the centre of the Earth. It becomes less with increasing distance from the centre of the Earth and varies according to the mass of the rocks below the surface. For geological purposes, the equivalent values of gravity (*g*) at sea level are calculated from the actual measurements made at a particular place. **gravitational** (*adj*).

**gravitational acceleration** the acceleration (p.151) due to gravity (↑); about 9.8 $m\ s^{-2}$.

**gal** (*n*) a unit for the measurement of gravity (↑); 1 gal is an acceleration (p.151) of 1 $cm\ s^{-2}$.

**milligal** (*n*) one-thousandth of a gal (↑); an acceleration (p.151) of 0.01 $mm\ s^{-2}$. The milligal is the unit generally used for measuring values of gravity in geophysical work.

**gravity meter, gravimeter** an instrument for measuring gravity (↑).

**gravity anomaly** the difference between the value of gravity (↑) measured at a particular place and the value for an imaginary Earth with no variations in density. Gravity anomalies can provide knowledge of variations in the density (p.154) of the rocks below the Earth's surface and are used in studying subsurface structures.

**isostacy** (*n*) the theory that the Earth's crust (p.9) is near to a state of equilibrium (p.155) without any tendency to move up or down, and that large blocks of the crust behave like blocks floating in a liquid. **isostatic** (*adj*).

**isostatic compensation** the means by which differences in the heights of parts of the Earth's crust (p.9) are balanced, either by 'roots' below them or by variations in the density (p.154) of the crust.

**isostatic adjustment** vertical movement in the Earth's crust (p.9) resulting from lack of isostatic equilibrium (↑), e.g. a rise in the level of the land surface after the weight of an ice sheet (p.28) has been taken away. The term 'isostatic adjustment' is also used to mean isostatic compensation (↑).

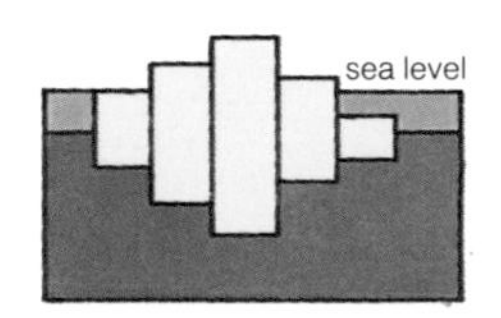


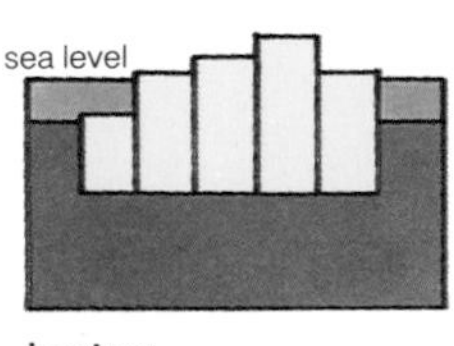


**isostacy**

**earthquake** (*n*) a sudden movement of part of the Earth's crust (p.9); a shock produced in the Earth's crust or mantle (p.9). An earthquake may be caused by movement along a fault (p.128) or by volcanic activity (p.68).

**seismology** (*n*) the study of earthquakes (↑). **seismological** (*adj*).

**seismic** (*adj*) relating to earthquakes (↑).

**seismograph** (*n*) an instrument for studying distant earthquakes (↑).

**focus** (*n*) the point from which an earthquake (↑) shock comes.

**epicentre** (*n*) the point on the Earth's surface directly above the focus (↑) of an earthquake (↑).

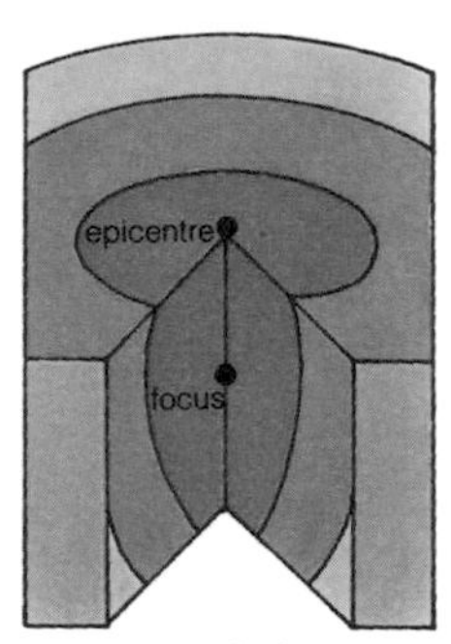

**earthquake focus and epicentre**

**magnitude** (*n*) a measure of the amount of energy (p.155) set free in an earthquake (↑). The magnitude of an earthquake is usually measured on the Richter scale.

**intensity** (*n*) a measure of the effects of an earthquake (↑) as estimated from the damage done. The scales of intensity generally used are the Mercalli, modified Mercalli, and Rossi–Forel.

**P-waves** earthquake (↑) waves in which the movements are in the same direction as that in which the waves travel. P-waves are of high frequency (p.156) and short wavelength (p.161).

**S-waves** earthquake (↑) waves in which the movements are at 90° to the direction in which the waves travel. S-waves are of high frequency (p.156) and short wavelength (p.161).

**Rayleigh waves** surface waves produced by an earthquake (↑) that give a rolling movement to the ground.

**L-waves** surface waves produced by an earthquake (↑) that cause horizontal movement at 90° to the direction in which the waves travel. L-waves are of low frequency (p.156) and long wavelength (p.161).

**isoseismal, isoseismal line** a line joining points at which the intensity (↑) of an earthquake (↑) is the same.

**shadow zone** the zone in which P-waves (↑) and S-waves (↑) are not received from a distant earthquake (↑). It lies between 103° and 143° from the epicentre (↑) of the earthquake.

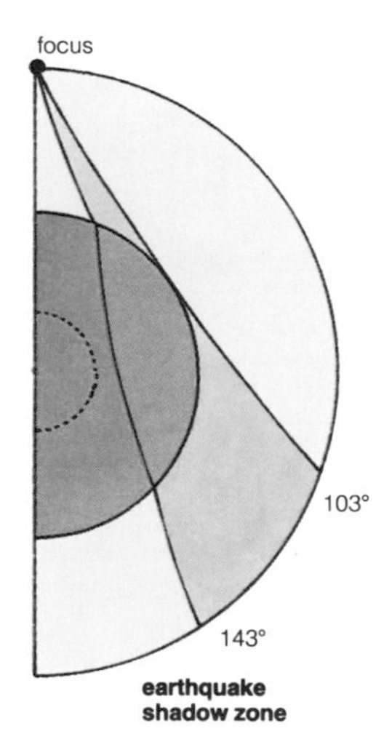

**earthquake shadow zone**

**microseism** (*n*) a very small movement of the Earth's crust (p.9) recorded by a seismograph (↑). Microseisms are caused by the wind, waves, etc.

**tremor** (*n*) a small earthquake (↑).

**principal shock** the main shock in a large earthquake (↑).

**fore-shock** a shock occurring before the principal shock (↑) in a large earthquake (↑).

**after-shock** a shock occurring after the principal shock (↑) in a large earthquake (↑).

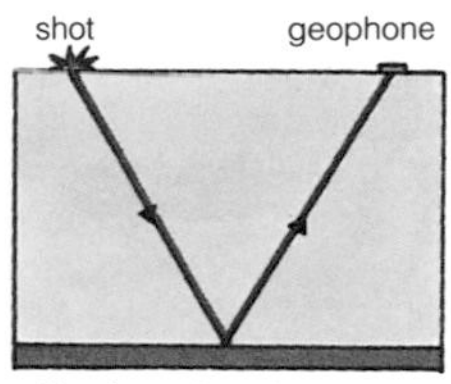

**reflection seismology**

**reflection seismology** a method of studying the rocks below the Earth's surface. Explosives are let off at the surface (a *shot*) to produce waves that pass down into the ground. *See diagram.* (A machine may be used instead of explosives.) The waves are *reflected*, i.e. thrown back, by discontinuities (p.10) below the surface and the reflected waves are detected at the surface by a device called a *geophone*. The time taken for the wave to travel from the surface to the discontinuity and back to the geophone is measured and can be used to calculate the depth of the reflecting layer. *See also* **refraction seismology** (↓).

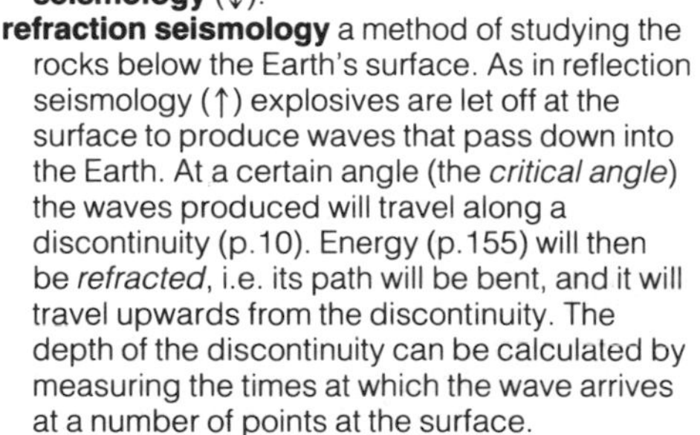
**refraction seismology** a method of studying the rocks below the Earth's surface. As in reflection seismology (↑) explosives are let off at the surface to produce waves that pass down into the Earth. At a certain angle (the *critical angle*) the waves produced will travel along a discontinuity (p.10). Energy (p.155) will then be *refracted*, i.e. its path will be bent, and it will travel upwards from the discontinuity. The depth of the discontinuity can be calculated by measuring the times at which the wave arrives at a number of points at the surface.

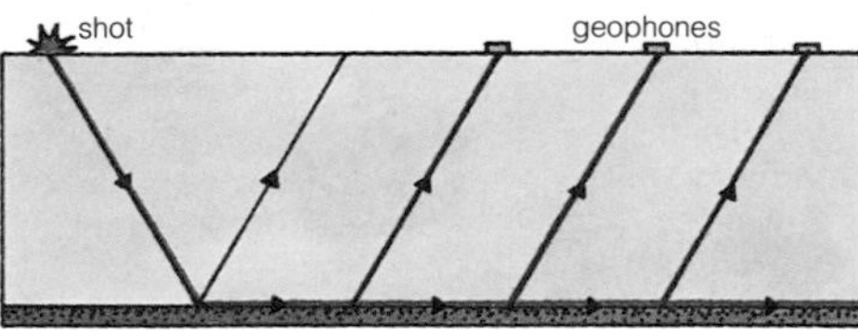

**refraction seismology**

**terrestrial magnetism** the magnetic field (p.157) of the Earth. The Earth's magnetic field is *dipolar*, i.e. it has two poles (north and south) like a bar magnet (p.157).

**geomagnetism** (*n*) (1) the Earth's magnetic field (p.157); (2) the study of the Earth's magnetic properties. **geomagnetic** (*adj*).

**magnetic poles** the two points on the Earth's surface to which a compass needle points. The Earth's north and south magnetic poles are not in the same places as the geographical poles and they move with time.

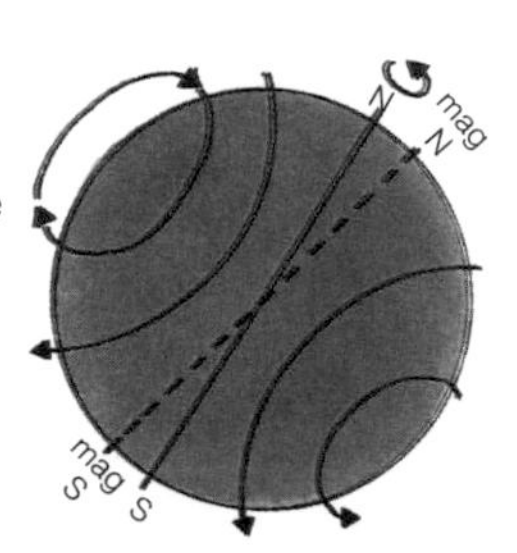


**magnetic poles**

**magnetometer** (*n*) an instrument for measuring the strength of the Earth's magnetic field.

**magnetic anomaly** (*n*) a variation or irregularity in the Earth's magnetic field (p.157) as measured at a particular place. Variations caused by differences in the magnetic properties of minerals and rocks are of value in studying the structure (p.122) of the rocks below the surface.

**palaeomagnetism** (*n*) 'fossil magnetism'; the magnetism of rocks that is thought to have been there since they were formed. Studies of the magnetization of rocks can provide information about the apparent movements of the Earth's magnetic poles (↑) that have taken place in the geological past and about the movements of continents. **palaeomagnetic** (*adj*).

**remanent magnetization** the lasting magnetization produced in a material (e.g. a rock) by a magnetic field (e.g. the Earth's magnetic field).

**Curie point** the temperature at which a material (e.g. a rock) loses any magnetization it has gained.

**reversed polarity** when the Earth's north and south magnetic poles (↑) change places the Earth's magnetic polarity is said to be *reversed*. This has happened many times in the geological past.

**palaeopole location** the geographical position of one of the Earth's magnetic poles (↑) at some time in the geological past, as shown by magnetic measurements made on rocks. *See also* **polar wander** (p.141).

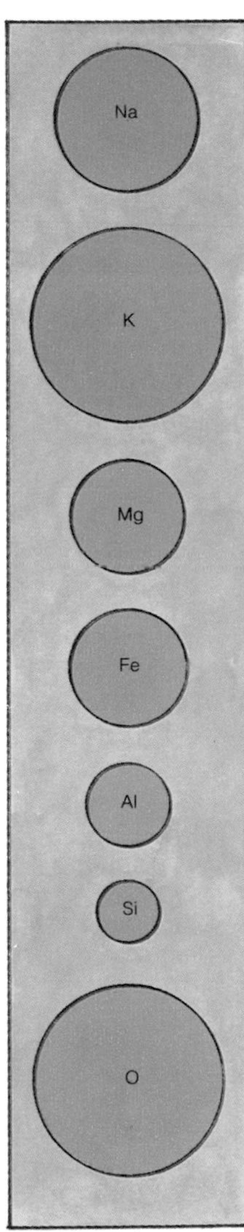

**Radii of some ions**

**geochemistry** (*n*) the chemistry (p.153) of the Earth, and especially the chemistry of the distribution of the elements (↓) in the various parts of the Earth. **geochemical** (*adj*).

**element** (*n*) a substance that cannot be broken down by chemical means into simpler substances.

**native element** an element (↑) occurring in nature in its free state as a mineral (p.44).

**trace element** an element (↑) occurring in very small quantities.

**rare earth element** one of a series of metallic elements (↑) which have very similar chemical properties. They occur in group III of the periodic table (p.18) and they all form a 'basic' oxide (an 'earth').

**REE** = rare earth element(s) (↑).

**ion** (*n*) an atom (p.152) that has gained or lost one or more electrons (p.155) and has become electrically charged (p.153).

**cation** (*n*) an ion (↑) with a positive electrical charge (p.153); e.g. a sodium ion is $Na^+$

**anion** (*n*) an ion (↑) with a negative electrical charge; e.g. $Cl^-$ (chlorine).

**ionic radius** (*radii*) a measure of the size of an ion (↑). The ionic radius, usually expressed in ångstrom units ($10^{-8}$ cm), is important in controlling the way in which ions will pack together.

**molecule** (*n*) the smallest particle of an element (↑) or chemical compound (↓) that can occur in the free state and has the properties of that element or compound. A molecule of an element consists of one or more atoms (p.152); a molecule of a chemical compound (↓) consists of one or more atoms of each of the elements that make up the compound. **molecular** (*adj*).

**molecular structure** the arrangement in space of the atoms (p.152) in a molecule (↑).

**chemical compound** a substance made up of two or more elements (↑) in a fixed proportion by weight.

**chemical composition** the nature of the elements (↑) that make up a substance and the proportions in which they are present in it.

**oxide** (*n*) a chemical compound (p.15) of oxygen and another element (p.15). Oxides may be solids, liquids, or gases. Solid oxides are present in the Earth's crust. Many ores (p.145) are oxides; e.g. haematite, $Fe_2O_3$ (p.48).
**alumina** (*n*) the oxide (↑) of aluminium, $Al_2O_3$.
**alkali** (*n*) (1) a chemical compound (p.15) that will dissolve (p.155) in water to produce a solution (p.159) that will neutralize (↓) an acid. (2) In geology, and especially in petrology (p.62), the word 'alkali' is used for rocks and minerals (p.44) that contain a large proportion of the alkali metals (↓). **alkaline, alkalic** (*adj*).
**halide** (*n*) one of the group of highly reactive (↓) elements (p.15) in Group VII of the periodic table (p.18): fluorine (F), chlorine (Cl), bromine (Br), and iodine (I). Halides occur naturally only as chemical compounds (p.15); e.g. sodium as the chloride (NaCl): halite (p.52).
**sulphide** (*n*) a chemical compound (p.15) of the non-metallic element sulphur (S) with another element, e.g. lead sulphide (PbS), the mineral galena (p.50).
**carbonate** (*n*) a chemical compound (p.15) containing the elements carbon (C) and oxygen (O) in the form of the $CO_3$ group containing one atom of carbon and three atoms of oxygen; e.g. calcium carbonate ($CaCO_3$), calcite (p.51).
**sulphate** (*n*) a chemical compound (p.15) containing the elements sulphur (S) and oxygen (O) in the form of the $SO_4$ group containing one atom of sulphur and four atoms of oxygen; e.g. barium sulphate, $BaSO_4$: barytes (p.52).
**phosphate** (*n*) a chemical compound (p.15) of the non-metallic element phosphorus (P) with oxygen (O) and another element; e.g. the mineral apatite (p.49).
**silica** (*n*) the oxide (↑) of silicon (Si), $SiO_2$. Quartz (p.55) is the most common natural form of silica.
**silicate** (*n*) a chemical compound (p.15) of silicon (Si), oxygen (O), and a metal or metals. The silicates are the most important group of rock-forming minerals (p.53).
**hydroxyl** (*n*) the group consisting of an oxygen atom joined to a hydrogen atom $-OH$.

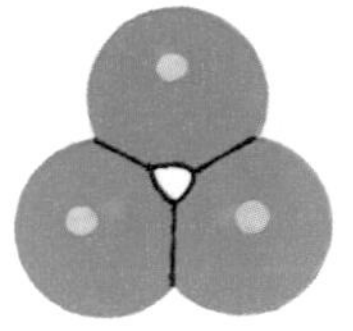

**carbonate ion $CO_3^{2-}$**

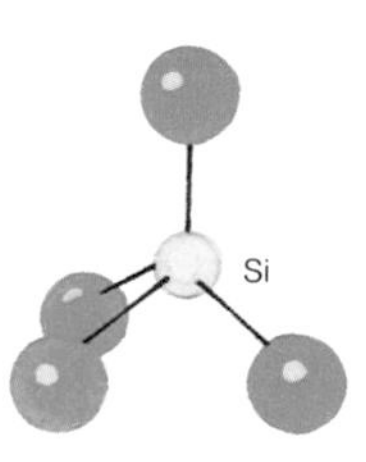


**$SiO_4$ tetrahedron**

the four oxygen atoms are at the corners of the tetrahedron:

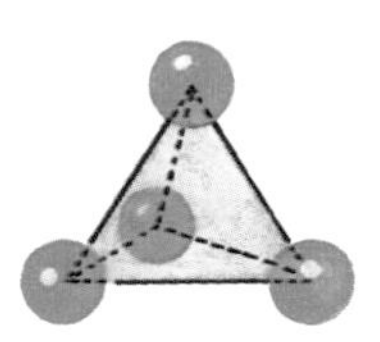

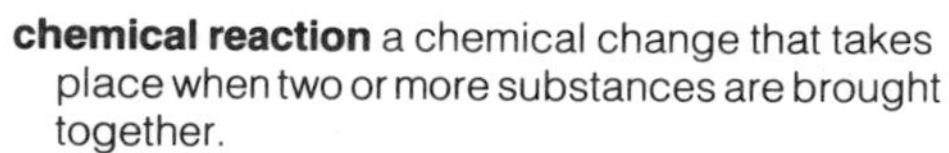



**chemical reaction** a chemical change that takes place when two or more substances are brought together.

**reactive** (*adj*) readily entering into chemical reactions (↑); chemically active. **react** (*v*); **unreactive** (*adj*), not reactive.

**chemical equilibrium** a state in which no further chemical reaction (↑) can take place in a chemical system.

**neutralization** (*n*) the chemical reaction (↑) between an acid and a base (p.152) in which a salt (↓) is formed. **neutralize** (*v*), **neutral** (*adj*).

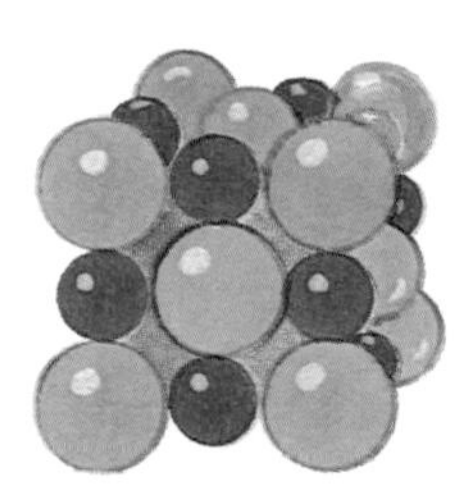

**sodium chloride NaCl structure**

**salt** (*n*) (1) a chemical compound (p.15) formed when the hydrogen of an acid is taken away and a metal is put in its place. *See also* **neutralization** (↑). (2) common salt, sodium chloride (NaCl).

**inorganic** (*adj*) obtained from minerals (p.44), i.e. from the Earth; not organic (↓).

**organic** (*adj*) describes chemical compounds (p.15) of carbon with hydrogen, some also containing oxygen, nitrogen, and other elements, found in living things. (Carbonates are not, however, regarded as organic compounds.)

**metal** (*n*) an element (p.15) whose atoms form positive ions (p.15) and which is generally a good conductor (p.153) of heat and electricity. **metallic** (*adj*).

**alkali metals** the metals (↑) lithium (Li), sodium (Na), potassium (K), rubidium (Rb), and caesium (Cs).

**non-metal** (*n*) an element (p.15) that does not have the general properties of a metal (↑). **non-metallic** (*adj*).

**analysis** (*n*) the use of chemical or physical (p.158) methods to find out the chemical composition (p.15) of a substance. **analyse** (*v*).

**alloy** (*n*) a material made up of two or more metals (↑) or a metal and a non-metal (↑). The composition of an alloy can vary slightly. Brass and steel are examples of alloys.

**NiFe** (*n*) an alloy (↑) of the metals (↑) nickel (Ni) and iron (Fe) that was thought to be the material of which the Earth's core (p.9) is made.

**geochemical cycle** the path followed by an element (p.15) during a series of geological changes, e.g. from magma (p.62) to rock and back to magma.

**lithophile** (*adj*) a lithophile element (p.15) is one that tends to collect in stony (silicate (p.16)) matter. Lithophile elements have a strong tendency to combine chemically with oxygen.

**siderophile** (*adj*) a siderophile element (p.15) is one that tends to collect in metallic iron. Siderophile elements have only a weak tendency to combine chemically with oxygen and sulphur.

**chalcophile** (*adj*) a chalcophile element (p.15) is one that has a strong tendency to combine with sulphur.

**abundances of elements** the relative amounts of various elements in the Earth, the Sun, and other stars can be calculated and expressed as abundances. *Terrestrial abundances* relate to the Earth; *cosmic abundances* to the stars.

**periodic table** the elements (p.15), when arranged in rows in a table in order of increasing atomic number (p.152), form groups in the columns of the table. The elements in any group have similar chemical properties (p.158).

**hydrolysis** (*n*) a chemical reaction (p.17) between water and another chemical compound (p.15). In geology, the word *hydrolysis* is used especially for processes in which minerals react with water, either as liquid water or as vapour (steam). For example, orthoclase (p.56) reacts with water to form kaolinite, a clay mineral (p.61). **hydrolyse** (*v*).

**volatile** (*n*) an element (p.15) or chemical compound (p.15) that is dissolved (p.155) in a magma at high temperature (p.160) or pressure. When the magma cools or the pressure is lowered, the volatiles come out of solution as gases. Water and carbon dioxide ($CO_2$) are examples. **volatile** (*adj*).

**precipation, chemical** the process by which a solid material, a **precipitate**, is produced in a liquid (usually water) by a chemical reaction (p.17). **precipitate** (*v*).

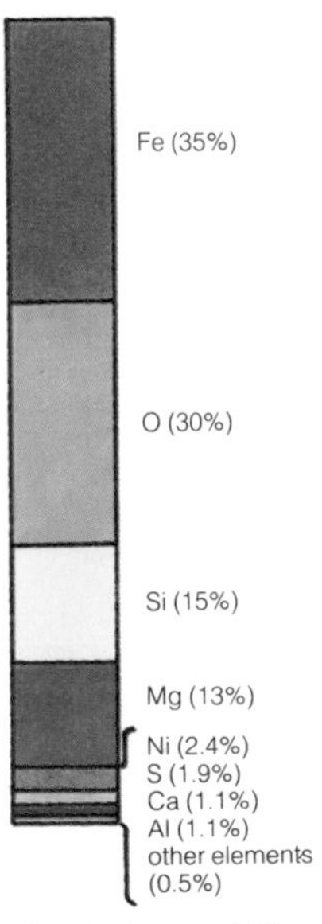


**chemical composition of the Earth**

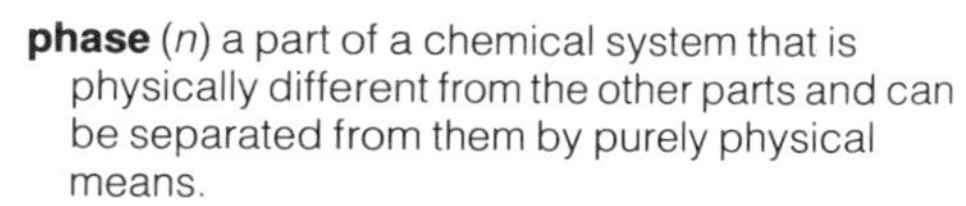

**phase** (*n*) a part of a chemical system that is physically different from the other parts and can be separated from them by purely physical means.

**component** (*n*) any single chemical element (p.15) or compound (p.15) in a chemical system of two or more components.

**open system** a chemical system that can gain or lose material or energy. *See also* **closed system** (↓).

**closed system** a chemical system that cannot gain or lose material or energy. *See also* **open system** (↑).

**isochemical** (*adj*) an *isochemical process* is one in which no material is brought in from outside.

**supercooling** (*n*) the cooling of a liquid below the temperature at which it would normally freeze. If a magma (p.62) is poured out at the Earth's surface and cools rapidly, the minerals formed may be different from those that would have appeared if the magma had crystallized slowly at depth. **supercooled** (*adj*).

**nucleation of crystals** if small crystal grains (p.72) are put into a supercooled (↑) liquid, crystallization (p.40) may begin. The grains act as *nuclei* (*sing. nucleus*) for crystallization.

**isotope** (*n*) certain elements have atoms (p.152) of more than one kind, called isotopes. All have the same atomic number (p.152) and almost the same chemical properties, but the atomic weights (p.152) of isotopes are different. **isotopic** (*adj*).

**radioactivity** (*n*) the property shown by certain elements (p.15) of changing into other elements by emitting (giving out) charged particles. **radioactive** (*adj*).

**half-life** the time taken for half the atoms in a piece of radioactive (↑) material to disintegrate (break up) into atoms of another element or isotope (↑).

**parent element** an element (p.15) that breaks down by radioactive (↑) decay to yield a daughter element (↓).

**daughter element** an element (p.15) produced by the radioactive (↑) decay (breakdown) of a parent element (↑).

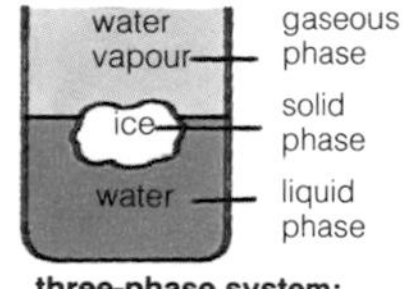


**three-phase system: one component**

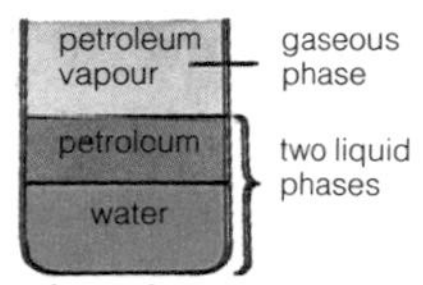


**three-phase system: two components**

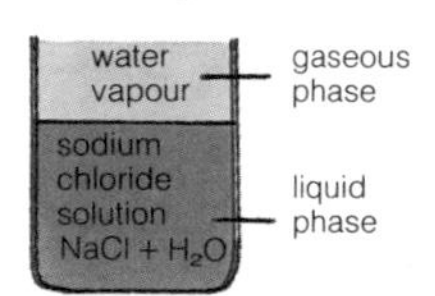


**two-phase system: two components**

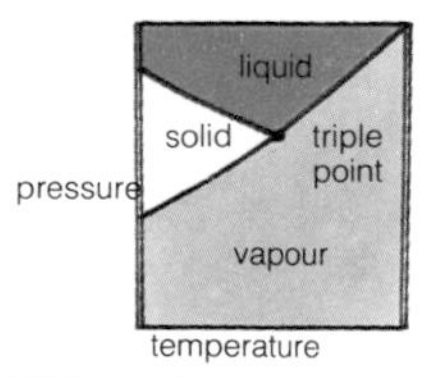


**triple point**

relationships between solid, liquid and vapour for water, three phases, solid, liquid and vapour, can exist together at the triple point

**weathering** (*n*) the process by which rocks at or near the surface of the Earth are broken up by the action of wind, rain, and changes in temperature. The effects of plants and animals are usually also included. Weathering is part of the process of erosion (↓). It includes mechanical weathering (↓) and chemical weathering (↓). **weather** (*v*), **weathered** (*adj*).

**unweathered** (*adj*) not weathered (↑).

**erosion** (*n*) (1) the wearing away of rocks: the effect of weathering (↑) and corrasion (p.24); (2) the processes by which soil and rock are removed from any part of the Earth's surface: part of the process of denudation (p.32), including weathering (↑), solution (p.159), corrasion (p.24), and transport (p.21). **erode** (*v*), **eroded** (*adj*).

**mechanical weathering** weathering (↑) produced by forces that break up the rock physically. These forces usually result from changes in temperature, e.g. insolation (↓); water freezing in cracks in a rock and forcing it apart; the growth of roots in cracks in the rock.

**insolation** (*n*) the effect of the sun's heat on rocks at the Earth's surface, especially the effect of changes in temperature on the mechanical weathering (↑) of rocks. Heating by the sun during the day causes rocks to expand (p.156). When they cool at night they contract (p.154). This causes the rock to break up. **insolate** (*v*).

**exfoliation** (*n*) the formation and breaking off of shells or sheets from the bare surfaces of rocks, especially granite and other igneous rocks (p.62). **exfoliate** (*v*).

**abrasion** (*n*) the wearing away of a rock by rubbing, e.g. by small particles of rock. **abrade** (*v*), **abraded** (*adj*).

**chemical weathering** weathering (↑) caused by chemical action, usually when water is present. For example, rain water containing carbon dioxide ($CO_2$) in solution (p.159) will dissolve (p.155) limestone (p.86).

**corrosion** (*n*) the eating away of rocks by chemical action. See also corrasion (p.24). **corrode** (*v*), **corroded** (*adj*).

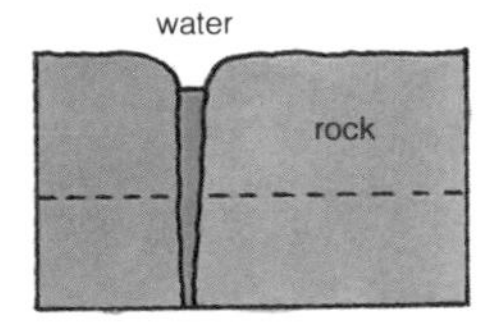


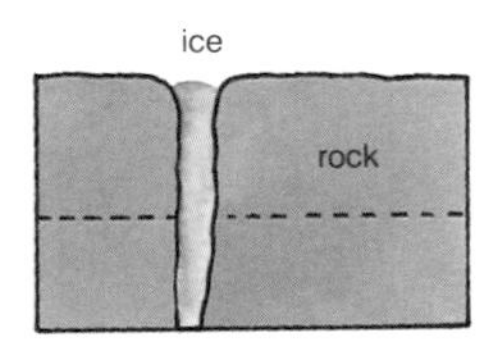


**mechanical weathering**
effect of water freezing in cracks

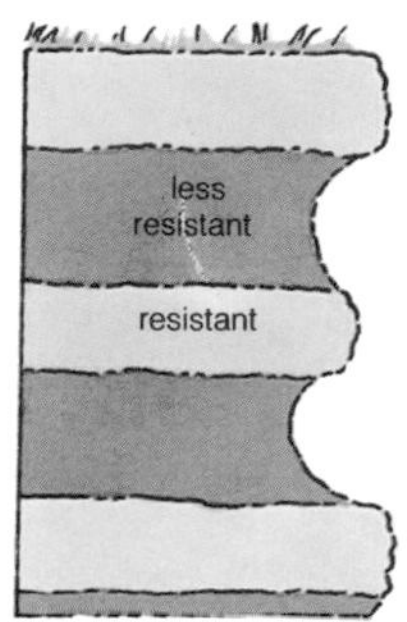

**differential weathering**

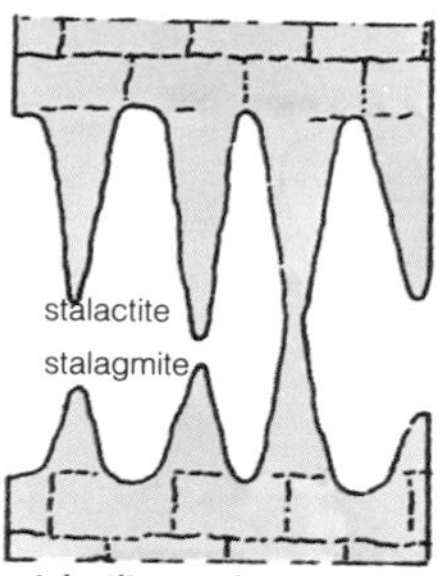

**stalactites, stalagmites**

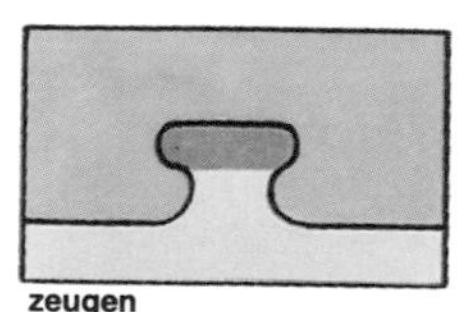
**zeugen**

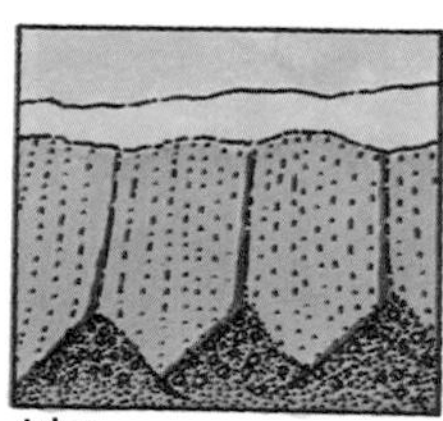
**talus**

**differential weathering, differential erosion** the process by which an uneven surface is developed where some rocks are worn away less rapidly than others and remain standing out in an exposure (p.122).

**joint** (*n*) a break or fracture (p.122) in a rock along which no movement has taken place.

**joint set** a series of joints (↑) that are more or less parallel to each other.

**joint system** two or more sets of joints (↑) that cut across each other.

**fissure** (*n*) a large crack or break in a body of rock. **fissured** (*adj*).

**tension gash** a joint (↑) that has opened up during deformation (p.122). It may contain minerals, e.g. quartz (p.55).

**sink-hole, swallow-hole** a hole in limestone (p.86) country into which water flows. Sink-holes are formed when the roof of a cave falls in or by the solution (p.159) of limestone at the surface.

**stalactite** (*n*) a deposit (p.80) of calcium carbonate, $CaCO_3$, hanging from the roof of a cave. *See also* **stalagmite** (↓).

**stalagmite** (*n*) a deposit of calcium carbonate, $CaCO_3$, standing up from the floor of a cave. *See also* **stalactite** (↑).

**calcareous tufa, calc tufa** a deposit (p.80) of calcium carbonate, $CaCO_3$, precipitated (p.18) from solutions (p.159). It is found in limestone (p.86) regions and round springs.

**zeugen** (*n*) a mass of harder rock resting on a pillar of softer rock.

**talus** (*n*) a loose heap of weathered (↑) pieces of rock at the foot of a steep slope. The pieces of rock may be of any size or shape.

**scree** (*n*) = talus (↑).

**angle of repose** the steepest angle or slope at which loose material such as pieces of rock or sand will remain without sliding down.

**rock glacier** a mass of pieces of rock flowing slowly down a slope like a glacier (p.28).

**transport, transportation** (*n*) the carrying-away of sediment (p.80) and other rock material on the surface of the Earth by gravity (p.11), moving water, ice, or air. **transport** (*v*).

**aeolian weathering** weathering (p.20) caused by the action of the wind. A strong wind blowing sand or other hard, sharp particles against a rock can wear it away.

**eolian weathering** (US) = aeolian weathering (↑).

**ventifact** (*n*) a stone shaped by the action of the wind.

**desert varnish** a thin, shiny coating that forms on stones in deserts. It is bluish-black in colour and consists mainly of iron and manganese oxides (p.16).

**dune** (*n*) a heap or bank of sand piled up by the wind into a regular shape.

**barchan dune** a dune (↑) shaped in plan like a crescent. The ends of the dune point in the direction in which the wind generally blows. Barchans may be up to 400 m wide and 30 m high.

**longitudinal dune** a long, narrow dune (↑), up to 80 km long and 200 m or more high, with its length parallel to the general direction in which the wind blows.

**seif** (*n*) a longitudinal dune (↑) with a long sharp edge at its top. One side of a seif is rounded; the other is a steep slip face. Seifs occur in chains.

**whaleback dune** a very large longitudinal dune (↑) with a flat top. There may be barchans (↑) or seifs (↑) on top of a whaleback dune.

**transverse dune** a dune (↑) with its length at 90° to the direction in which the wind generally blows.

**parabolic dune** a dune (↑) shaped in plan like the path of a ball thrown in the air (i.e. a *parabola*). The points of the dune face the direction from which the wind blows. Parabolic dunes are formed where there is thick grass or other plants covering the sand. The sand is blown away from an area without plants but the sand on either side is held back by the plants there. A parabolic dune is thus formed.

**oghurd** (*n*) a large mountainous dune (↑).

**loess** (*n*) an unconsolidated (p.84) deposit (p.80) of silt (p.88), usually unstratified (p.80), carried by the wind.

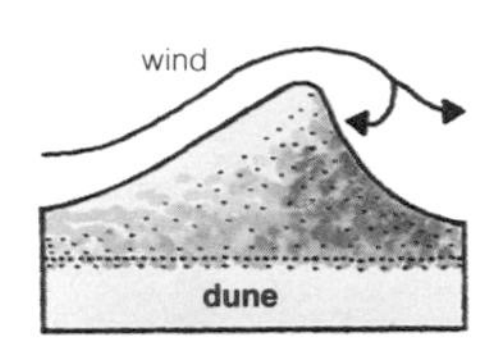


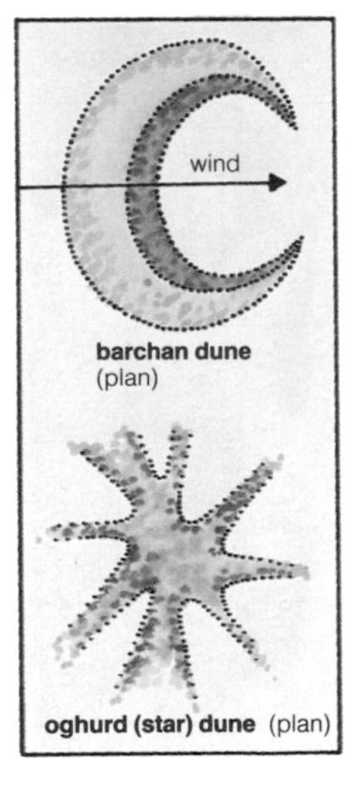


**barchan dune** (plan)

**oghurd (star) dune** (plan)

**soil** (*n*) the material produced by the effects of weathering and the action of plants and animals on the rocks at the Earth's surface.

**subsoil** (*n*) the layer of broken rock between the soil (↑) and the unweathered solid rock, or *bedrock*, below.

**pedology** (*n*) the study of soils; soil science.

**regolith** (*n*) a layer of pieces of loose rock resting on solid rock (bedrock).

**mantle rock** = regolith (↑).

**soil profile** the three soil horizons (↓) A, B, and C.

**soil horizon** one of the layers into which the soil is divided.

**A horizon** the top layer of the soil. It is dark in colour and contains organic (p.17) material formed by the decay of vegetable matter.

**B horizon** the subsoil; the layer below the soil surface. It usually contains more clay and iron oxides (p.16) than the A and C horizons.(↑ ↓).

**C horizon** the lowest layer of the soil, resting on unweathered (p.20) solid rock. It consists of loose, slightly weathered pieces of rock.

**Cca horizon** a white layer of calcium carbonate, $CaCO_3$, beneath the B horizon (↑).

**K horizon** a thick hard layer containing 50% or more of calcium carbonate, $CaCO_3$, that forms beneath the B horizon (↑) in very dry regions.

**caliche** = K horizon (↑).

**A2 horizon** a whitish layer between the A and B horizons (↑). Most of the iron oxides (p.16) have been removed from the A2 horizon by water moving down through the soil.

**duricrust** (*n*) a hardened layer formed in the soils of very dry (semi-arid) regions by the precipitation (p.18) of salts (p.17) from water in the soil.

**bauxite** (*n*) a residual (p.33) deposit (p.80) formed under very hot, wet conditions. It contains hydrated (p.157) aluminium oxides (p.16). Bauxite is an important ore (p.145) of aluminium.

**laterite** (*n*) a residual (p.33) deposit (p.80) formed under very hot, wet conditions, especially from igneous rocks (p.62). It contains hydrated (p.157) iron oxides (p.16). **laterization** (*n*).

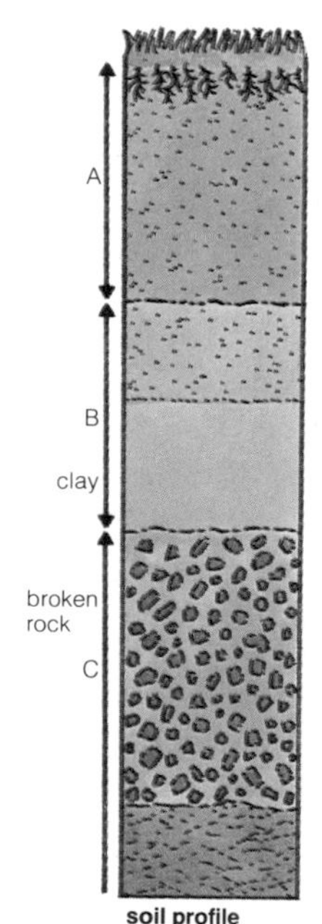

**soil profile**
(residual soil)

**load** (*n*) the material moved by a stream. It may float in the water, be pushed along the stream bed, or be dissolved (p.155) in the water.
**suspension load** the part of the load (↑) of a stream that is carried along in the water above the stream bed. *See also* **traction load** (↓).

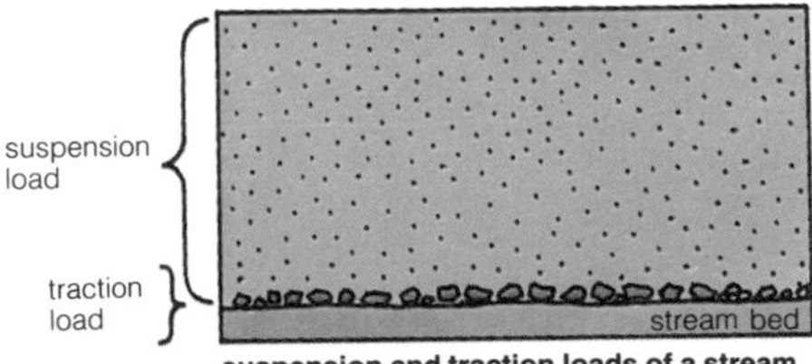

**suspension and traction loads of a stream**

**traction load** the part of the load (↑) of a stream that is carried along on the stream bed. *See also* **suspension load** (↑).
**corrasion** (*n*) the wearing away of rocks by the rubbing action of particles carried by a stream.
**alluvium** (*n*) mud, sand, gravel (p.87), and other materials moved by streams and deposited (p.80) by them. **alluvial** (*adj*).
**alluvial fan** a cone-shaped pile of alluvium (↑) deposited (p.80) where the gradient (↓) of a stream becomes less (e.g. at the base of a steep slope). Alluvial fans are common in dry regions.
**stream gradient** the slope of a stream bed as measured down the valley.
**longitudinal profile, long profile** a curved line representing the way in which the height of the valley floor changes along the course of a stream.
**Thalweg** (*n*) = longitudinal profile (↑).
**stream profile** = longitudinal profile (↑).
**rejuvenation** (*n*) if a region is uplifted (p.125) or the sea-level falls, the streams are *rejuvenated* and cut down into the land again. **rejuvenate** (*v*).
**knick-point** (*n*) the point at which the old longitudinal profile (↑) of a stream meets a new one. A knick-point is the result of rejuvenation (↑).
**base level** the imaginary level surface to which the longitudinal profile (↑) of a stream is related.
**mature** (*adj*) a mature stream is one that has reached its full growth. **maturity** (*n*).

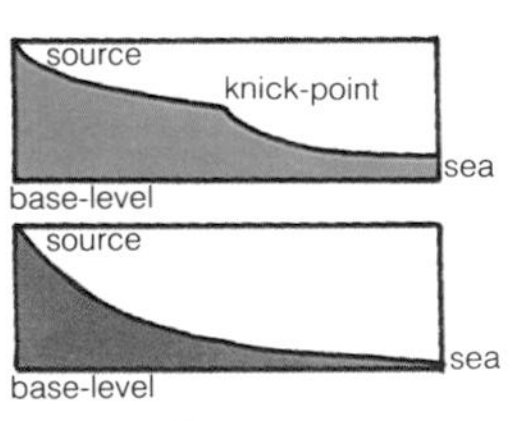

**stream profiles (exaggerated)**

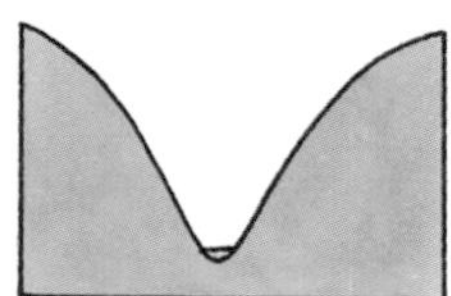
V-shaped valley

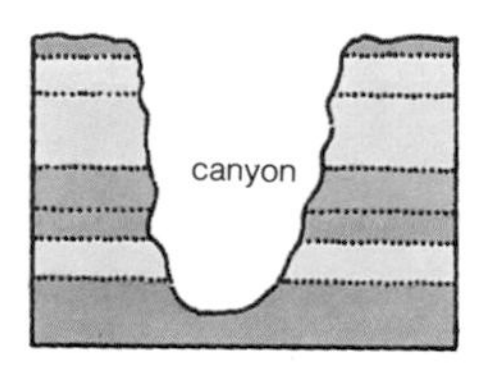


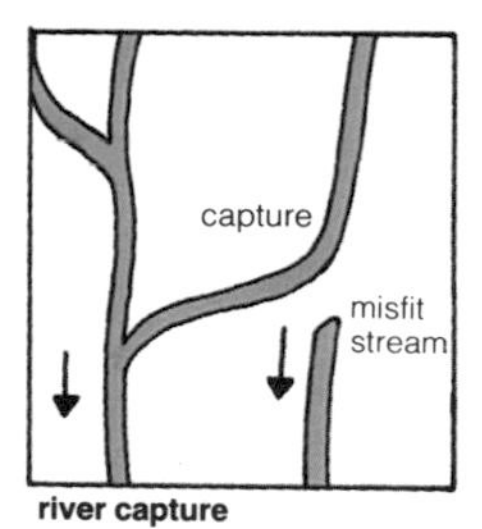

river capture

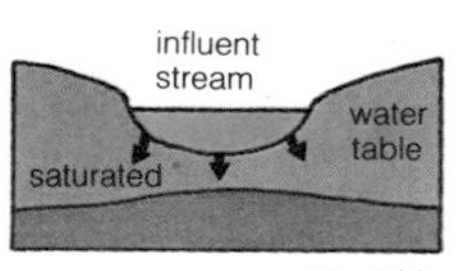


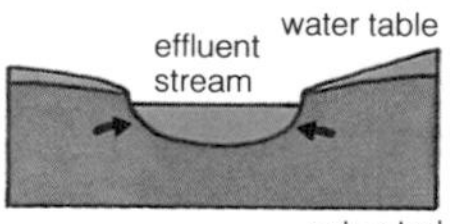


**V-shaped valley** a valley with steep sides and in cross-section like a letter V. V-shaped valleys are characteristic of young streams.

**canyon** (*n*) a deep valley with sides that are vertical, or nearly so, which has been cut by a river. Canyons are usually formed by rejuvenation (↑).

**waterfall** (*n*) a point in the course of a stream where the water descends more or less vertically without support.

**river terrace** a flat area that borders a river valley. It slopes towards the sea at about the same angle as the river. A terrace marks the level of the floor of an earlier valley.

**capture** (*n*) a stream that is actively eroding (p.20) may cut back and reach the upper part (the headwaters) of another stream, thus capturing these waters and turning them aside into its own course. **capture** (*v*).

**beheaded** (*adj*) a beheaded stream is one that has had its upper part (its headwaters) captured (↑) by another stream. **behead** (*v*).

**misfit stream** a stream that is too small to have eroded (p.20) the valley in which it flows. River capture (↑) is a possible cause.

**watershed** (*n*) (1) the line that divides two areas from which water flows into two separate streams; a narrow area of ground between two such areas; (2) the area from which water flows into a particular stream system.

**drainage system** a stream or river together with the streams that flow into it.

**influent** (*adj*) an influent stream is one that flows above the water table (p.146) and thus adds to the supply of water below ground. Influent streams are common in dry regions. *See also* **effluent** (↓).

**effluent** (*adj*) an effluent stream is one that flows at the level of the water table (p.146) and receives water from it. *See also* **influent** (↑).

**dry valley** (*n*) a valley without a stream. Stream capture is one cause of dry valleys.

**wadi** (*n*) a valley in which a stream flows from time to time. Wadis are common in deserts.

**arroyo** (*n*) = wadi (↑).

**flood-plain** the flat area on either side of a stream over which it spreads when too much water is flowing for the stream channel to be able to carry all of it.

**meander** (*n*) the curved path of a river, especially in a flood-plain (↑). **meandering** (*adj*), **meander** (*v*).

**incised meander** a meander (↑) that has been cut down (incised) in the flood-plain (↑). The river then flows in a twisting channel with steep sides.

**ox-bow lake** a lake shaped like a new moon that has been formed when a meander (↑) has been cut off from the main stream by continuing erosion.

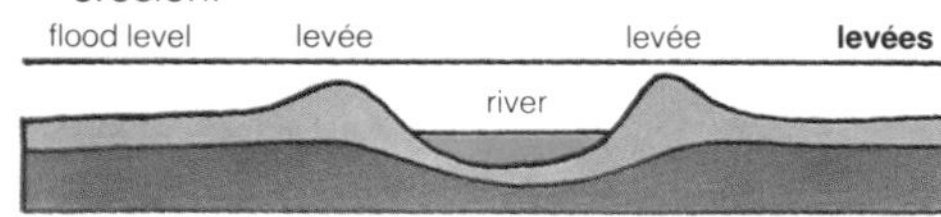

**levées**

**levée** (*n*) a natural wall or embankment that has been formed of sediment (p.80) deposited at the sides of a river when it has flowed over its banks. Levées are usually found in the flood-plain (↑) and they contain the river while the flow of water is not too large.

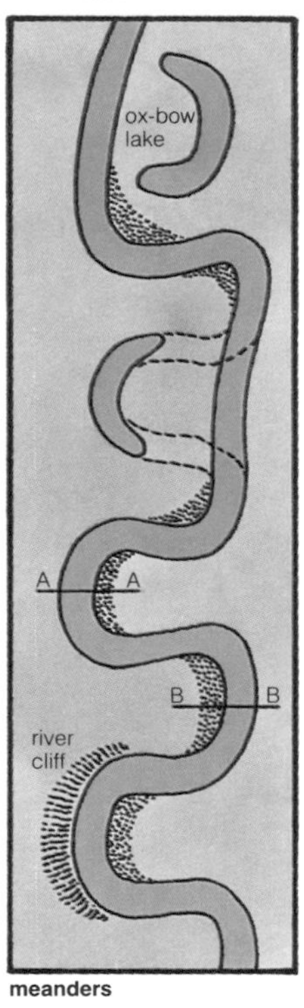

**meanders**

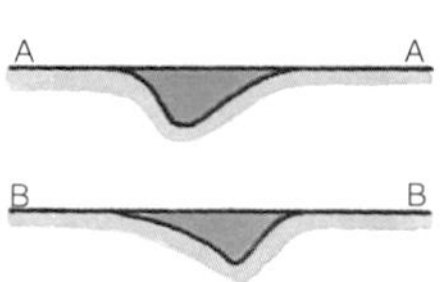


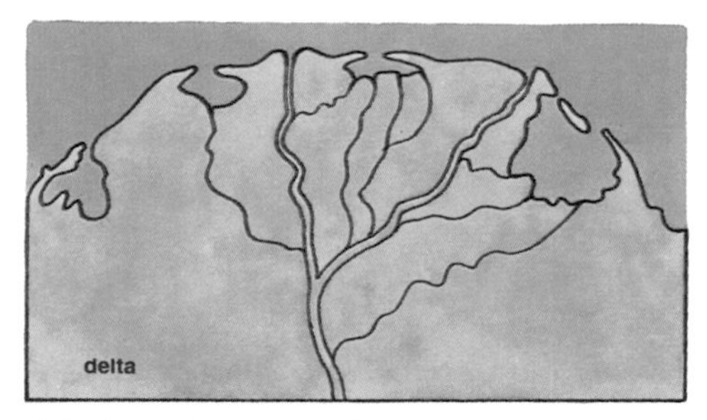


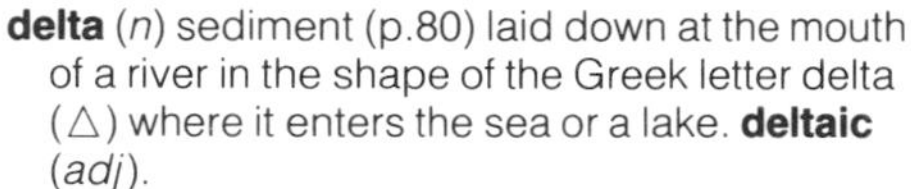

**delta** (*n*) sediment (p.80) laid down at the mouth of a river in the shape of the Greek letter delta (△) where it enters the sea or a lake. **deltaic** (*adj*).

**distributary** (*n*) one of the branches into which a river divides in a delta (↑) or elsewhere.

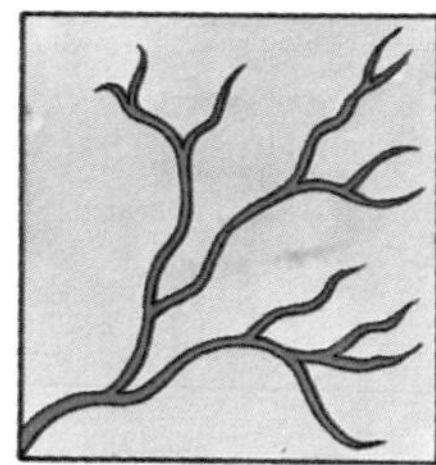
dendritic

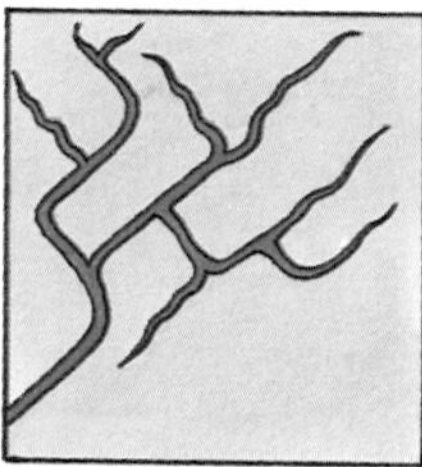
rectangular

radial

**drainage patterns**

**drainage pattern** the arrangement of a system of streams as seen in plan. Common types of drainage patterns are: *dendritic*, in which the streams are arranged like the branches of a tree; *trellis* or *rectangular*, in which the streams flow in two directions at 90° to each other, parallel to the strike and dip (p.123) of the rocks; and *radial*, in which the streams flow out in all directions from a central point.

**consequent, subsequent and obsequent streams**

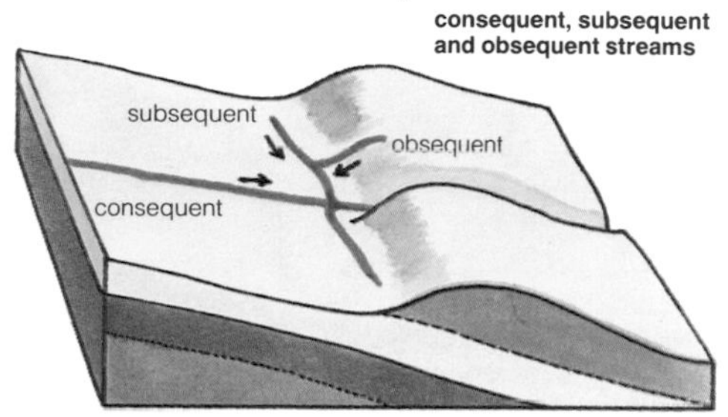


**consequent** (*adj*) a consequent stream is one that flows in the same direction as the downward slope on which it was originally formed.

**subsequent** (*adj*) a subsequent stream is one that flows into a consequent (↑) stream.

**obsequent** (*adj*) an obsequent stream is one that flows in the opposite direction to the original consequent stream (↑).

**inconsequent** (*adj*) an inconsequent drainage system is one that does not fit the geological structure (p.122) of the region.

**superimposed** (*adj*) superimposed drainage is a form of inconsequent drainage (↑). It results when a drainage system is formed on a younger series (p.113) of rocks that rest with an angular unconformity (p.118) on older rocks. The drainage is fitted to the younger series of rocks and is then superimposed when the streams cut down through them to the older rocks, whose structure is not related to the drainage system.

**antecedent** (*adj*) an antecedent stream or drainage system cuts across a geological structure (p.122) that has formed across its course.

**ice sheet** a mass of ice in the form of a sheet covering a large area of the Earth's surface. Ice sheets can stretch across continents and can cover mountains.

**glacier** (*n*) a large mass of ice formed from snow that has packed together and moves slowly down a slope under its own weight.

**mountain glacier** a glacier (↑) that flows in a mountain valley with solid rocks standing above the highest levels of ice and snow.

**valley glacier** = mountain glacier (↑).

**alpine glacier** = mountain glacier (↑).

**piedmont glacier** a glacier that is formed where a mountain glacier (↑) leaves the valley and spreads out across a plain.

**crevasse** (*n*) a deep crack in a glacier. Crevasses are caused by stresses (p.122) in the ice that result from movement.

**Bergschrund** (*n*) a crevasse (↑) at the head of a glacier (↑) in a cirque (p.31).

**glaciation** (*n*) the action of glaciers (↑) and ice-sheets (↑), including erosion (p.20) and deposition (p.80).

**glaciology** (*n*) the study of glaciers (↑) and ice-sheets (↑). **glaciological** (*adj*).

**glacier lake** a lake, usually in a valley, in which the water is held back by a glacier (↑).

**ice-dammed lake** = glacier lake (↑).

**glacial lake** = glacier lake (↑).

**glacial period** a period of time in the Earth's history when ice in the form of glaciers (↑) spread into areas that were free from it at other times.

**ice age** = glacial period (↑).

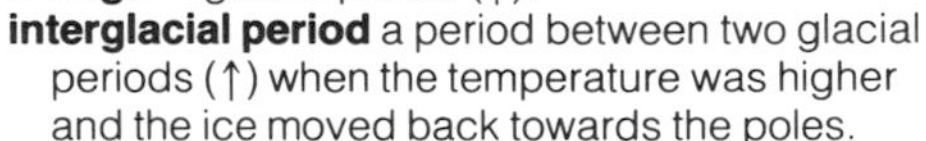

**interglacial period** a period between two glacial periods (↑) when the temperature was higher and the ice moved back towards the poles.

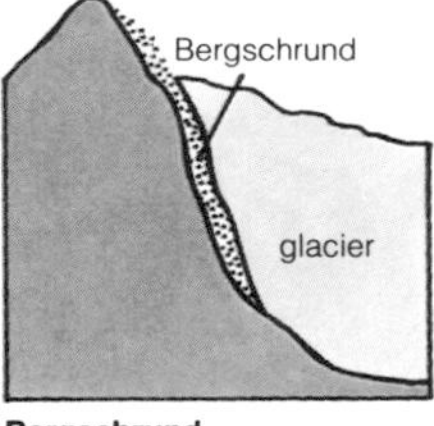


**Bergschrund**

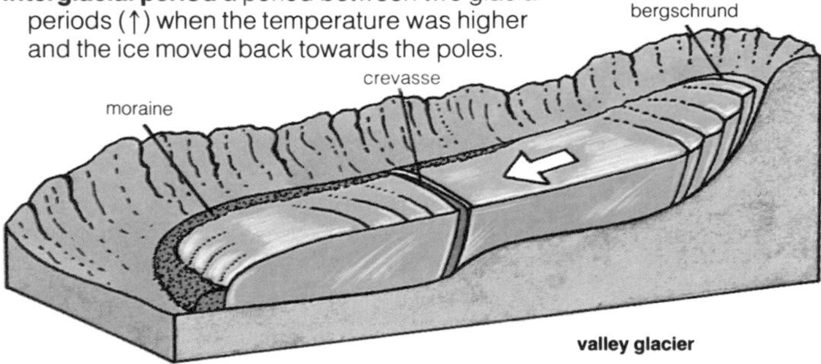


**valley glacier**

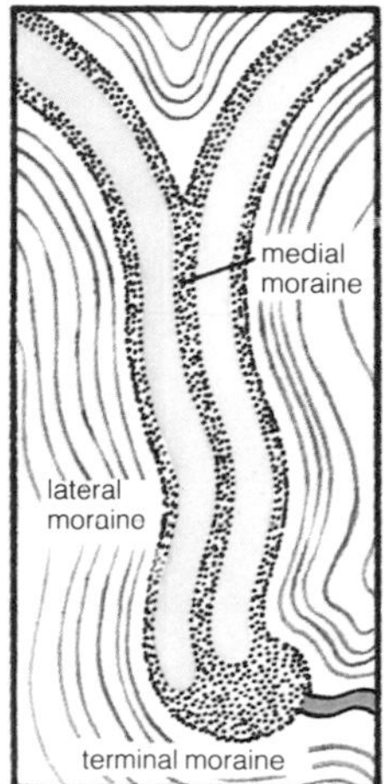
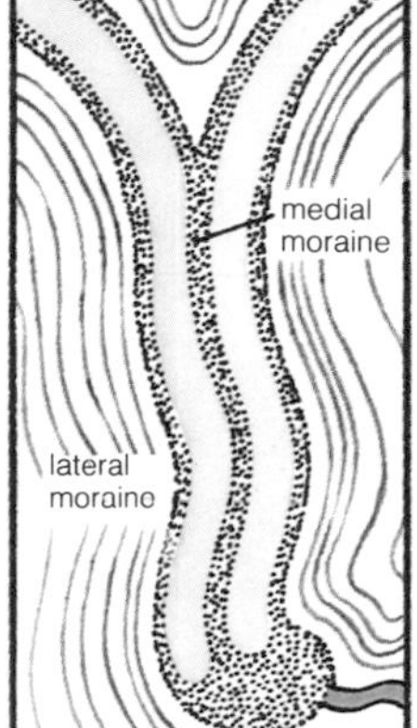


**glacial moraines**

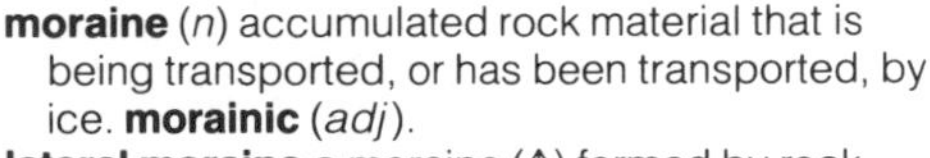

**moraine** (*n*) accumulated rock material that is being transported, or has been transported, by ice. **morainic** (*adj*).

**lateral moraine** a moraine (↑) formed by rock material falling on the sides of a glacier (↑) from the sides of the valley.

**medial moraine** a moraine (↑) formed by the joining together of lateral moraines (↑) when two glaciers (↑) meet in the same valley.

**englacial moraine** a moraine inside the ice of a glacier (↑).

**terminal moraine** a moraine (↑) left where the ice melts at the lower end, or *snout*, of a glacier (↑).

**end-moraine** = terminal moraine (↑).

**boulder clay** a deposit (p.80) left behind after the melting of an ice-sheet (↑): a fine clay containing pebbles (p.87) and boulders (p.87) of subangular (p.83) shape.

**till** (*n*) = boulder clay (↑).

**tillite** (*n*) an indurated (p.84) till (↑).

**glacial striae** long marks made on rocks that have been under a glacier (↑). Striae are made by pieces of rock that are carried along with the ice.

**glacial striations** = glacial striae (↑).

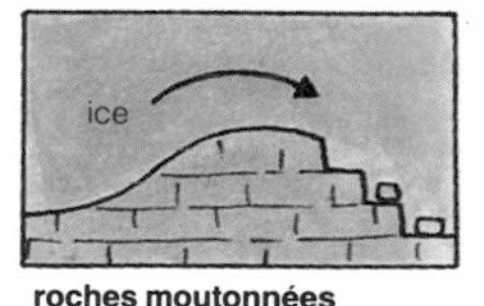


**roches moutonnées**

**roches moutonnées** (*n.pl., French*) rock forms produced by glacial erosion (p.20). They have a rounded shape, less steep on the side from which the ice travelled, and are usually oval in plan. Roches moutonnées are generally found in groups with a parallel arrangement. **roche moutonnée** (*sing.*).

**permafrost** (*n*) ground that is frozen all the time.
**frost heave** the process by which soil and rocks are lifted up by the freezing of water below the surface.
**frost wedging** the action of frost in forcing rocks apart.
**solifluction** (*n*) the flow of wet material at the surface that takes place when the ground in a permafrost (↑) area is partly unfrozen.
**pingo** (*n*) a small hill formed by ice action in a permafrost (↑) region.
**esker** (*n*) a long, narrow hill of gravel and sand in a region that was once covered by ice. Eskers usually follow a twisting course, with many bends. They are probably formed by water from streams flowing through the ice. *See* **kame** (↓).

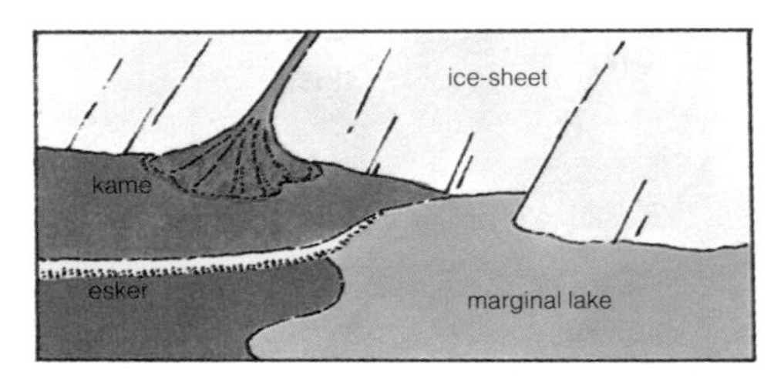


**kame** (*n*) a hill, usually conical in shape, made of glacial deposits (p.80) and formed at the edge of an ice-sheet (p.28) by water flowing from the ice. The word 'kame' is also used with various other meanings.
**drumlin** (*n*) an oval mound of boulder clay (p.29). Drumlins usually occur in groups.
**marginal lake** a lake formed at the edge of an ice-sheet (p.28).
**glaciofluvial deposits** deposits (p.80) formed by water that comes from a glacier or ice-sheet (p.28).
**fluvio-glacial deposits** = glaciofluvial deposits (↑).
**outwash fan** glaciofluvial deposits (↑) that have formed at the front of an ice-sheet or large glacier (p.28).

**outwash plain** an area of more or less level ground at the edge of a glacier (p.28) where glaciofluvial deposits (↑) are formed.

**glacial erratic** a large piece of rock, or boulder, that has been carried by ice for some distance and has then come to rest where the ice has melted.

**drift** (*n*) glacial (p.28) or glaciofluvial (↑) deposits.

**periglacial** (*adj*) on the borders of an ice-sheet (p.28). The word 'periglacial' is used both for the geographical area and for the physical conditions in it.

**U-shaped valley** a valley that in cross-section is shaped like a letter U. The shape is produced by glacial (p.28) erosion (p.20).

**hanging valley** a smaller valley that joins a larger valley high above the floor of the larger valley. Hanging valleys are usually the result of glacial (p.28) erosion (p.20).

**hanging valley**

**truncated spur** a hill at the side of a valley whose end has been cut off by the action of a glacier (p.28).

**fjord, fiord** (*n*) a glaciated (p.28) valley of which part is below the sea.

**cirque** (*n*) a hollow of rounded shape at the head of a mountain valley. Cirques are formed by ice action.

**corrie** (*n*) = cirque (↑).

**cwm** (*n*) = cirque (↑).

**arête** (*n*) a sharp-edged ridge formed by glacial (p.28) erosion (p.20), commonly between two cirques (↑).

**geomorphology** (*n*) the study of land forms; the study of the surface forms of the Earth and their development. **geomorphological** (*adj*).

**erosional cycle** the idea that the erosion (p.20) of a region can be understood as a process that repeats itself again and again – a cycle. Each cycle begins with uplift (p.125) of the land; hills and valleys (relief) (↓) are then formed. At the end of the cycle the hills are worn down and a level plain, a *peneplain* (↓), is left. Rivers are the means by which the erosional cycle is brought about in parts of the world where the weather is neither very hot nor very cold. In regions where the rainfall is very small the cycle ends with a pediplain (↓), which may be covered by a thin layer of alluvium (p.24).

**cycle of erosion** = erosional cycle (↑).

**denudation** (*n*) the lowering of the land surface by all the processes of erosion (p.20); the laying bare of the rocks by the carrying away of the material covering them. **denude** (*v*), **denuded** (*adj*).

**degradation** (*n*) the wearing down of the rocks to a lower level. **degrade** (*v*), **degraded** (*adj*).

**aggradation** (*n*) the opposite of degradation (↑): the building up of a surface by deposition (p.80). **aggrade** (*v*).

**gradation** (*n*) the combined effects of aggradation (↑) and degradation (↑) on the Earth's crust. Gradation can be divided into three processes: the erosion (p.20) of the surface; the transport (p.21) of the eroded material; the deposition (p.80) of the eroded material.

**peneplain, peneplane** (*n*) 'almost a plain'. A land surface that has been eroded to a nearly level plain.

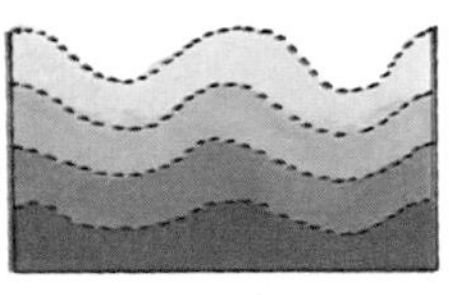

**development of a peneplain**

**pediplain** (*n*) a kind of peneplain (↑). An erosion surface of large area formed by the joining up of two or more pediments (↓).

**relief** (*n*) the shape of part of the Earth's surface as it is shown in differences of height and steepness of slopes.

**plateau** (*n*) (*plateaux*) a flat or nearly flat area of high ground with steep sides that stands above the country round about it.

**cuesta** (*n*) a long, narrow hill with a steep slope (*escarpment* (↓)) on one side and a gentle slope (*dip-slope* (↓)) on the other.

**escarpment** (*n*) a steep slope or cliff on one side of a cuesta (↑).

**dip-slope** (*n*) the gentle slope on one side of a cuesta (↑), corresponding to the dip (p.123) of the rocks.

**pediment** (*n*) a gently sloping surface produced by the erosion of cliffs or steep slopes. A pediment is usually cut in solid rock with a thin layer of sediment (p.80) resting on it.

**residual** (*adj*) remaining above the general level of an area of land that has been worn down to a plain; as, e.g. a monadnock (↓) or inselberg (↓).

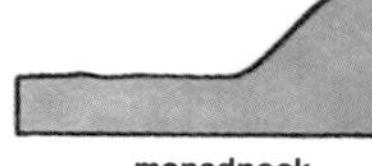

monadnock

**monadnock** (*n*) a residual (↑) hill or mountain standing by itself above a peneplain (↑).

**inselberg** (*n*) an 'island mountain'. A residual (↑) hill or mountain with steep sides and a round top that stands by itself in a plain. A type of monadnock (↑).

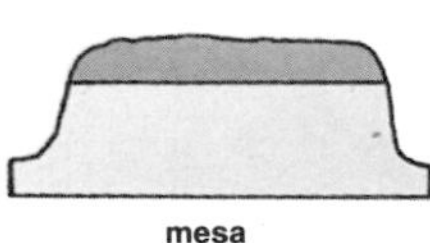

mesa

**mesa** (*n*) an area of high, flat land (tableland) with steep sides. It has horizontal beds of hard rock at the top.

**butte** (*n*) a small flat-topped hill formed by the erosion (p.20) of a mesa (↑).

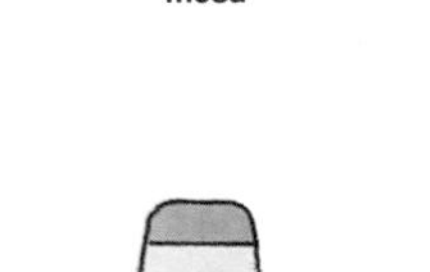

butte

**karst topography** an uneven type of countryside found in limestone (p.86) areas. Groundwater (p.146) makes its way through the rocks and dissolves (p.155) them, and streams flow below the surface.

**térra rossa** 'red earth'. A residual (↑) deposit (p.80) formed in karst topography (↑) by the solution (p.159) of limestone (p.86).

**hydrosphere** (*n*) all the water in, on, or above the Earth's surface: in the oceans, rivers, or lakes, under the ground or in the air.
**oceanography** (*n*) the study of the oceans. **oceanographic, oceanographical** (*adj*).
**marine** (*adj*) of the sea, in the sea, or formed by the sea.
**continental shelf** the nearly level part of the sea floor, next to a land mass, over which the sea is not more than 180–200 m deep.
**shelf** (*n*) short for *continental shelf* (↑).
**neritic** (*adj*) relating to shallow seas, especially to the waters over the continental shelf (↑).
**neritic zone** the part of the sea floor from low tide mark (p.37) to the outer edge of the continental shelf (↑).
**epicontinental sea** a shallow sea over part of a continental shelf (↑).
**continental slope** the sloping part of the sea floor from the outer edge of the continental shelf (↑). It slopes at 3–6°.
**bathyal** (*adj*) relating to the sea and the sea floor of the region of the continental slope (↑) from about 200 m to 2000 m depth.
**marginal plateau** (*plateaux*) a level part of the sea floor at a greater depth than the continental shelf – usually at depths between 240 m and 2400 m.
**abyssal** (*adj*) of the deep ocean and its floor; at more than 2000 m depth.
**abyssal plain** a level area of the floor of the deep ocean.
**abyssal hills** relatively small hills that rise up from the floor of the deep ocean to heights of about 1000 m.
**continental rise** the gently sloping surface between the continental slope (↑) and the abyssal plain (↑).

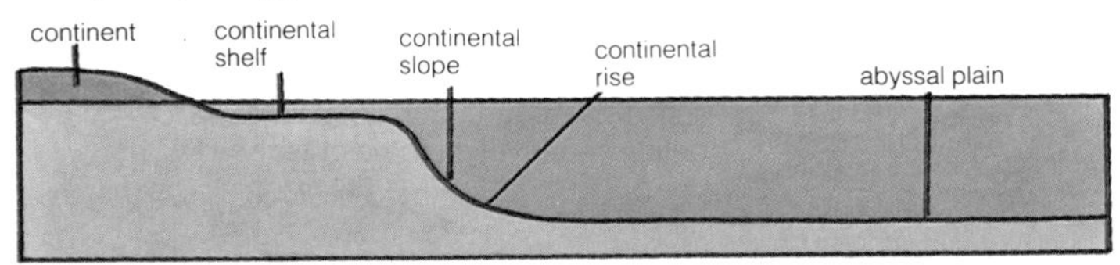


**the oceans**

**submarine** (*adj*) below the sea.
**submarine canyon** a valley with steep sides cut in the continental shelf (↑) or continental slope (↑). Some submarine canyons also cut into the continental rise (↑).
**ocean trench** a long, narrow area, much deeper than the rest of the ocean floor, that runs parallel to an island arc (p.135) or a continental margin (p.157).
**oceanic ridge** a mountain range under the ocean that is long and relatively narrow.
**mid-oceanic ridge** an oceanic ridge (↑) in the middle of an ocean, e.g. the mid-Atlantic Ridge.

**mid-oceanic ridge in section**

**submarine plateau** a generally flat area that is higher than the rest of the ocean floor.
**seamount** (*n*) a mountain under the sea rising from the floor of the ocean but not reaching sea-level; most are of volcanic (p.68) origin.
**guyot** (*n*) a seamount (↑) with a flat top. Most guyots are thought to be volcanoes (p.68) that have been eroded by wave action.

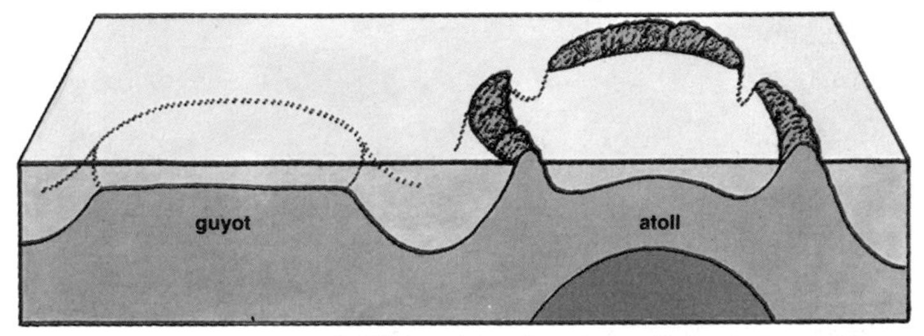


**atoll** (*n*) an island or chain of islands formed from coral reefs (p.38) shaped like a ring round an area of water (a lagoon).
**isobath** (*n*) a line joining points on the sea-bed (or other surface) that are all at the same depth below sea-level or some other horizontal surface: a submarine contour.

**wave base** the depth in the sea below which there is no effective erosion (p.20) or transport (p.21) of material by waves.

**turbidity current** a mass of water carrying sediment (p.80) that travels with violent movement down a slope under water. The sediment it carries makes the turbidity current denser than the seawater around it, and its higher density causes it to move down the slope. The sediment deposited by a turbidity current is called a **turbidite**.

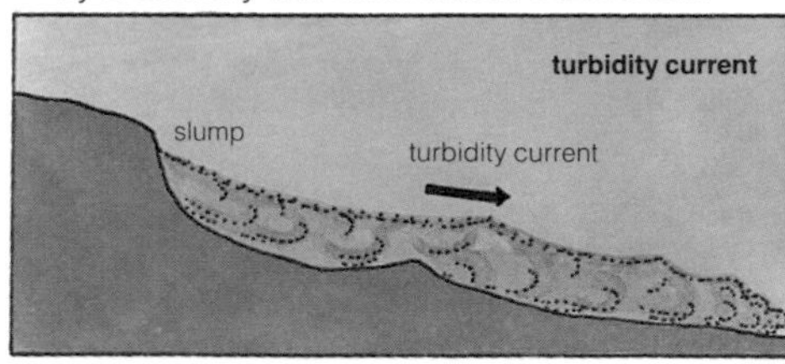


**density current** = turbidity current (↑).

**tsunami** (*n*) a sea wave caused by an earthquake under the sea.

**eustatic** (*adj*) *eustatic* movements are world-wide changes in sea level.

**photic zone** the part of the sea that the light of the sun reaches: down to about 200 m.

**ooze** (*n*) a deposit (p.80) in the abyssal zone (p.34) of the oceans. Oozes contain the hard parts (skeletons) of tiny animals living in the sea. There are two types of ooze: *calcareous* (p.86) *oozes*, formed at depths from 2000 m to 3900 m, and *siliceous* (p.87) *oozes*, formed at depths greater than 3900 m.

**Globigerina ooze** a calcareous (p.86) deposit (p.80) formed on the floor of the deep ocean. It is made up largely of the sheels of Foraminifera (p.104), especially *Globigerina*.

**Radiolarian ooze** a siliceous (p.87) deposit (p.80) formed on the floor of the deep ocean. It is made up largely of the skeletons of Radiolaria (p.104).

**red clay** a deposit (p.80) formed in the deepest parts of the oceans, at depths of more than 5000 m. It is made up of fine material carried by the wind, including volcanic (p.68) dust.

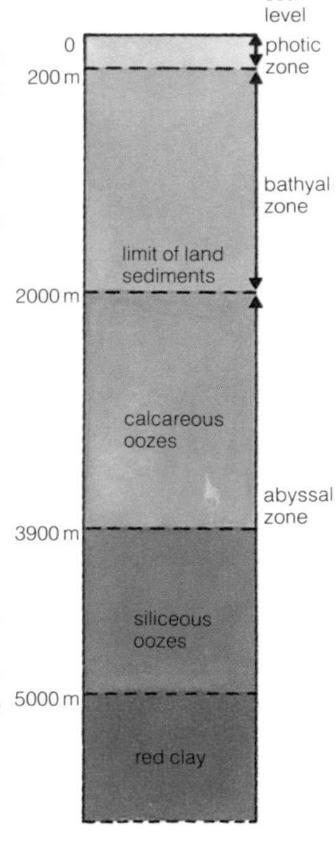

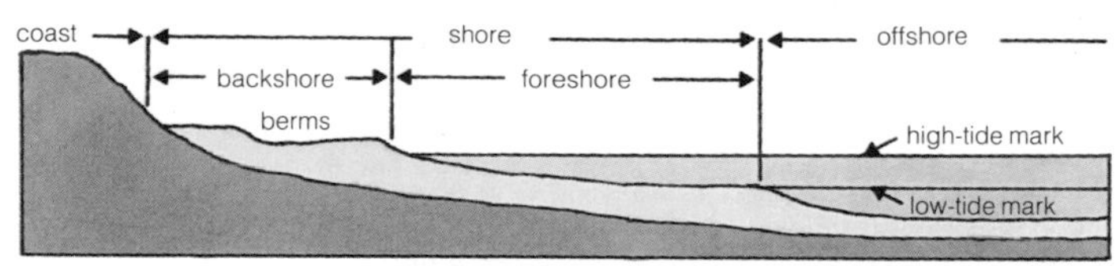


**beach**

**coast** (*n*) the part of the land that is next to the sea and on which waves have a direct effect.
**coastline** (*n*) (1) the line that divides the coast (↑) from the shore (↓); (2) the line that divides the land from the sea. *See also* **shoreline** (↓).
**tide** (*n*) the rise and fall of the water in the oceans, caused by the gravitational (p.11) pull of the sun and moon. **tidal** (*adj*).
**tide mark** the point on a shore (↓) or cliff reached by the water of the sea at high or low tide (↑).
**shore** (*n*) the region from low tide mark (↑) to the highest point reached by waves or tides (↑).
**shoreline** (*n*) the line at which the land meets the sea. *See also* **coastline** (↑).
**off-shore** (*adj*) from low tide mark (↑) to the edge of the continental shelf (p.34).
**berm** (*n*) the backshore: the part of the shore that is level or slopes towards the land.
**backshore** (*n*) = berm.
**foreshore** (*n*) the zone between low tide mark (↑) and the berm or backshore (↑).
**littoral** (*adj*) of the shore (↑); between high and low tide marks.
**beach** (*n*) the shore (↑), especially where it is formed of sand or pebbles (p.87).
**shingle** (*n*) small rounded stones and loose pebbles (p.87); material characteristic of beaches (↑).
**cliff** (*n*) a high and very steep rock face, nearly vertical, especially one on the sea shore (↑).
**sea cave** a cave that has its entrance in a sea cliff (↑). A sea cave is likely to be formed where a cliff is made of stratified sediments (pp.80, 81).
**wave-cut platform** a level or nearly level area of rock formed by wave action below a sea-cliff. It may be bare or covered by a beach (↑).
**notch, wave-cut notch** a V-shaped cut at the base of a sea-cliff formed by wave action.

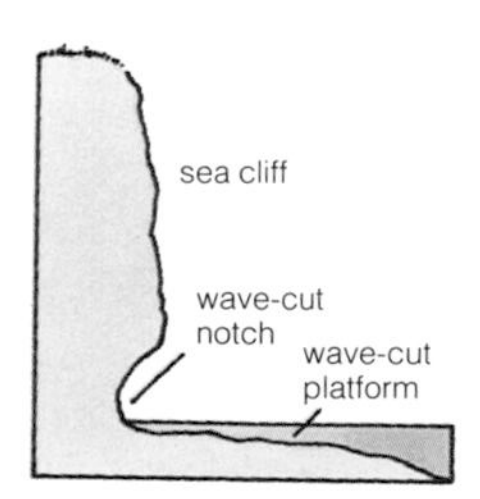

**stack** (*n*) a mass of rock with steep sides standing off shore (p.37) by itself like a small island. Stacks are formed when a sea cliff (p.37) is eroded (p.20).

**arch** (*n*) a mass of rock in the shape of an arch. Arches are commonly formed by coastal erosion (p.20) of stratified rocks (p.80) where harder beds (p.80) rest on weaker beds.

**spit** (*n*) a long, narrow deposit (p.80) of sediment (p.80) that stands out from a coast.

**bar** (*n*) a long, narrow deposit (p.80) of sand or gravel (p.87) that stands at or not far below sea level, either at the mouth of a river or parallel to a beach (p.37).

**barrier island** a long island or beach (p.37) parallel to a shoreline. There is usually a lagoon (↓) between the barrier island and the coast. The island is usually only one or two metres above sea level. A barrier island may be formed by the growth of a spit (↑) or by the submergence (↓) of a coast after dunes (p.22) have been formed.

**reef** (*n*) a narrow ridge of rock at or near the surface of the water.

**coral reef** a reef (↑) made up of corals (p.105) or other organisms; a type of bioherm (p.101).

**barrier reef** a long, narrow coral reef (↑), parallel to the shore and separated from it by a lagoon (↓).

**lagoon** (*n*) (1) a body of shallow water between a barrier island (↑) and the shore; (2) a body of water inside the coral (p.105) reefs (↑) of an atoll (p.35). **lagoonal** (*adj*).

**sabhka** (*n*) a broad, flat coastal area with a thin covering of salt (NaCl). Sabhkas lie above the high tide mark (p.37) and are not often covered by the sea.

**marine swamp** a wet area on a coast. The water moves slowly if at all. Plants grow in large numbers and plant remains make up a large part of a marine swamp. In the geological past marine swamps were common in the Carboniferous period (p.114). They can be seen today on the eastern and southern coasts of the United States.

stack

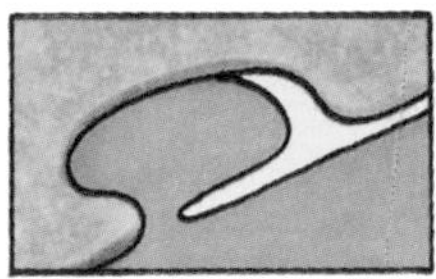
bar

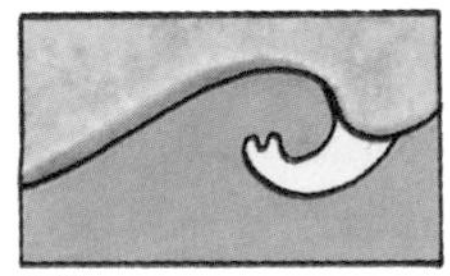
spit

**raised beach**

**primary coast** a coast (p.37) formed by the sea coming to rest against a land form that has been shaped by terrestrial (p.81) activities: erosion, deposition, volcanism, fault movements. *See also* **secondary coast** (↓).

**secondary coast** a coast (p.37) that has been shaped by the sea or by marine (p.34) organisms (p.98). *See also* **primary coast** (↑).

**submerged coast** a coast (p.37) that is the result of a rise in sea level, or of downward movement of the land. The sea usually reaches inland in embayments (↓), which may go far in from the coast.

**drowned coast** = submerged coast (↑).

**embayment** (*n*) an inward curve in a shoreline (p.37) forming an open *bay*.

**emergent coast** a coast (p.37) that is the result of a fall in sea level or of the rising of the land. Emergent coasts commonly have marine terraces (↓) with beaches (p.37), wave-cut platforms (p.37), and old sea cliffs (p.37) standing above sea level.

**raised beach** a beach that is above sea level and is separated from the present beach. Raised beaches are characteristic of emergent coasts (↑).

**marine terrace** an old marine (p.34) beach (p.37) or wave-cut platform (p.37) now standing some distance above sea level.

**ria** (*n*) an embayment (↑) or arm of the sea that has been shaped by stream erosion (p.20) before being filled by the sea.

**ria coast** a submerged coast (↑) in which valleys formed by stream erosion (p.20) have been filled by the sea.

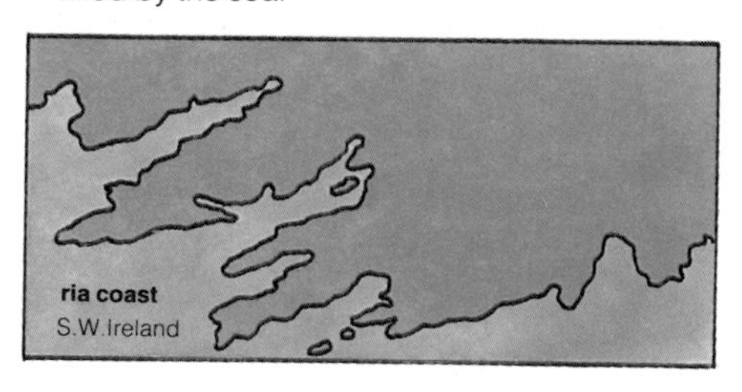
**ria coast**
S.W.Ireland

**crystal** (*n*) a body with surfaces that are smooth, flat, and regularly arranged. The regular shape of a crystal results from the regular arrangement of the atoms (p.152) of which it is made. **crystalline** (*adj*); *see also p.44.*

**crystallize** (*v*) to form crystals (↑). **crystallization** (*n*).

**crystallography** (*n*) the study of crystals (↑). **crystallographic** (*adj*).

**crystal lattice** the regular arrangement of atoms (p.152) in three dimensions in a crystalline solid.

**unit cell** the smallest complete piece of a crystal lattice (↑) that shows the arrangement of the atoms (p.152) in a crystal. The unit cell contains a number of atoms arranged in a regular way. It is repeated in three dimensions to form the crystal lattice.

**face** (*n*) a single, flat surface on a crystal (↑). In crystallography it is not the sizes of faces that are important but the angles between them – the interfacial angles (↓).

**interfacial angle** the angle between two faces (↑) of a crystal (↑). It is measured between lines at 90° to the crystal faces. These lines are called *normals*.

**normal** (*n*) *see* interfacial angle (↑).

**goniometer** (*n*) an instrument for measuring the interfacial angles (↑) of crystals (↑). There are two types: the *contact goniometer*, in which two straight arms are placed on the crystal faces to be measured; and the *reflecting goniometer*, in which a beam of light is used to make the measurement.

**form** (*n*) in crystallography, a form is a group of crystal faces that are related to a single face by the symmetry elements (p.42) of a particular crystal class. For example, eight faces make up a pyramid form in the cubic or tetragonal system (p.43); six faces make up the form of a hexagonal prism. A closed form (e.g. a pyramid) can enclose space by itself; an open form (e.g. a prism) cannot.

**zone** (*n*) a set of crystal faces (↑) that meets at edges parallel to each other is called a *zone*. **zonal** (*adj*).

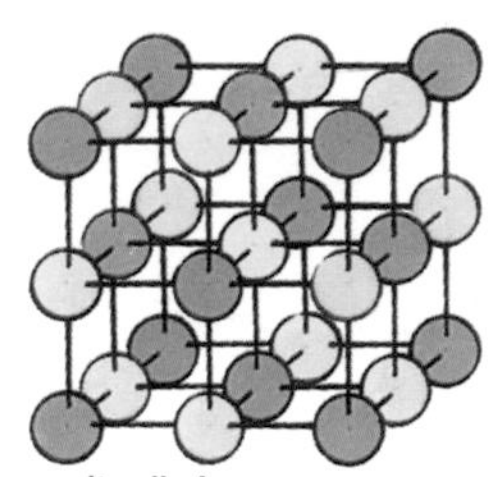

**unit cell of sodium chloride (NaCl)**

each $Na^+$ ion is surrounded by six $Cl^-$ ions and each $Cl^-$ ion by six $Na^+$ ions

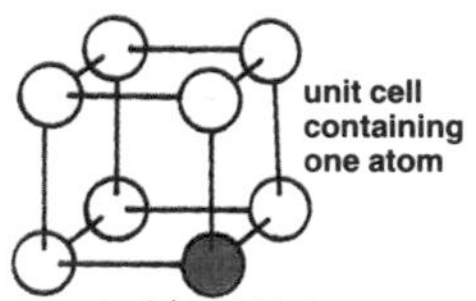

**unit cell containing one atom**

each of the eight atoms shown is common to eight unit cells

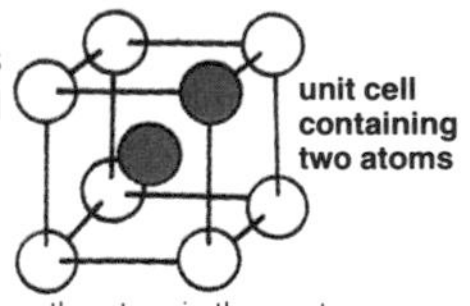

**unit cell containing two atoms**

the atom in the centre belongs only to the unit cell shown

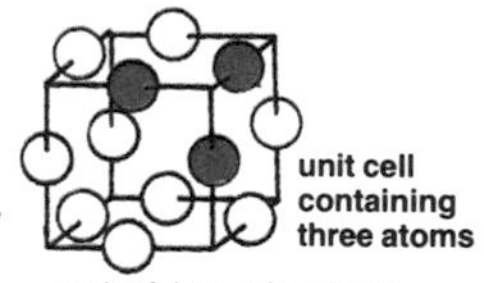

**unit cell containing three atoms**

each of the twelve atoms shown belongs to four unit cells

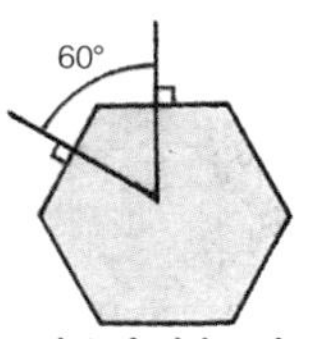


**interfacial angle**

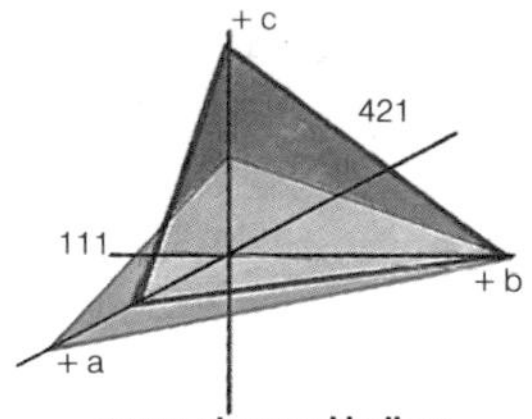


**parameters and indices**
blue: unit form, 111
red: plane with parameters ½, 1, 2; indices 421

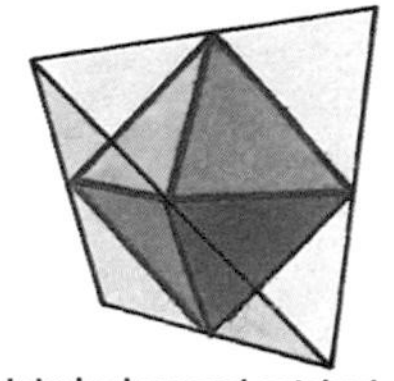

**tetrahedron and octahedron**
(grey) (red)

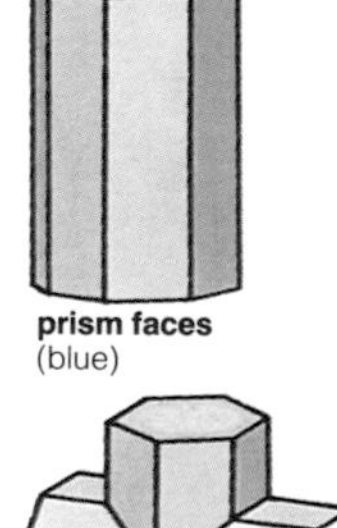

**prism faces**
(blue)

**twin crystal**

**crystallographic axis** (*axes*) one of the set of three or four imaginary lines used for measuring and describing the forms of crystals (↑). **axial** (*adj*).

**intercept** (*n*) the distance between the point where a plane, such as a crystal face (↑), cuts a crystallographic axis (↑) and the point where the axes meet each other (the origin).

**parameter** (*n*) the ratios of the intercepts (↑) that the unit form (↓) makes with the crystallographic axes (↑) are called the *parameters*.

**axial ratio** = parameter (↑).

**unit form** a crystal form (↑) is chosen as the unit form to provide the parameters (↑) for measuring other forms present in the same crystal.

**indices** (*n.pl.*) the crystallographic indices are obtained by calculating in turn the value of the *reciprocal* of each parameter (↑), i.e. the value of 1 divided by the parameter, and then turning the figures obtained into whole numbers. The indices are used for describing crystals.

**cube** (*n*) a solid with six faces (↑), each of which is a square, at 90° to each other.

**tetrahedron** (*n*) (*tetrahedra*) a solid with four faces (↑), each of which is an equilateral triangle (a triangle with all sides equal in length). **tetrahedral** (*adj*).

**octahedron** (*n*) (*octahedra*) a solid with eight faces (↑), each of which is an equilateral triangle (a triangle with all sides equal in length). **octahedral** (*adj*).

**prism** (*n*) a crystal face (↑) that cuts the horizontal crystallographic axes (↑) and is parallel to the vertical (*c*) axis. **prism, prismatic** (*adj*).

**pyramid** (*n*) an open crystal form (↑) consisting of faces that meet at a point.

**pinacoid** (*n*) an open crystal form (↑) consisting of two faces parallel to each other.

**twin crystal** two crystals (↑) of the same substance joined together in such a way that a crystallographic (↑) direction or crystallographic plane is shared by the two parts. Twins may be *simple, penetrating* (in which one crystal appears to pass through the other), *repeated*, or *compound* (*complex*). **twinning** (*n*), **twinned** (*adj*).

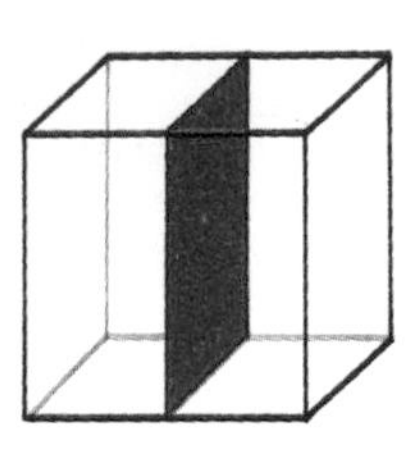

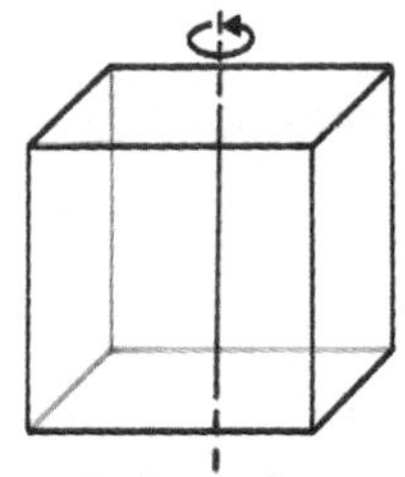

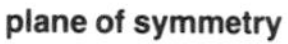

**plane of symmetry**

**axis of symmetry**

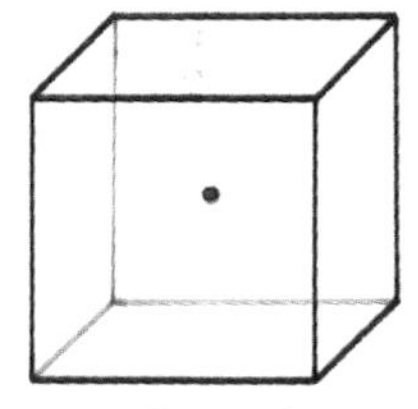

**centre of symmetry**

**symmetry** (*n*) in crystallography (p.40), the exact agreement of faces (p.40) on opposite sides of a crystal (p.40). Crystallographic symmetry depends on the angles between the faces (p.40) of a crystal.

**symmetry element** a plane (↓), axis (↓), or centre (↓) of symmetry.

**plane of symmetry** an imaginary flat surface that divides a body, such as a crystal (p.40), into two halves, each of which is like the other as reflected in a mirror.

**axis of symmetry** an imaginary line on which a crystal may be turned so that it comes into positions that are crystallographically the same two or more times in one complete turn of 360°. An axis of symmetry can be twofold, threefold, fourfold, or sixfold (but not fivefold).

**centre of symmetry** a centre of symmetry is present when for each point on the surface of a crystal there is a similar point on the opposite side of the crystal, the two points being on a straight line passing through the centre of the crystal and being at equal distances from the centre.

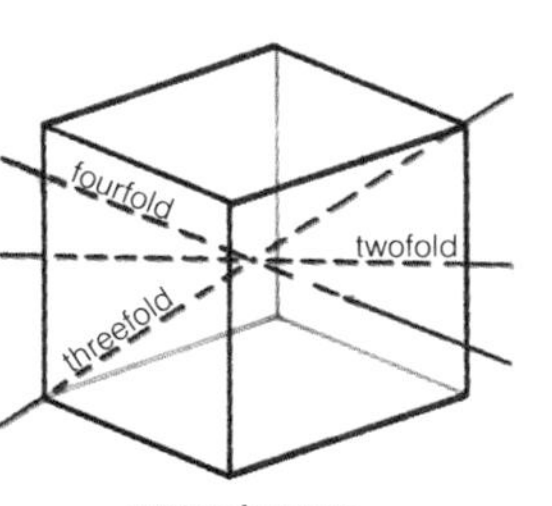


**symmetry axes of the cube**

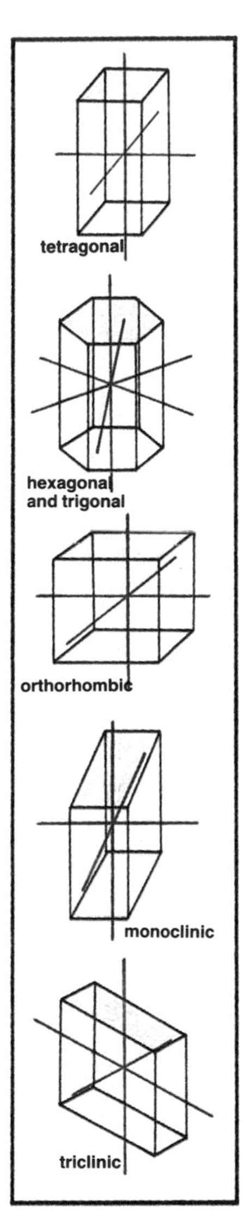


**crystal system** any crystal (p.40) can be classified as belonging to one of seven crystal systems, each of which has its own symmetry elements (↑). The same set of crystallographic axes (p.41) is used for describing all the crystals belonging to any one system.

**cubic system** a crystal system (↑) with four threefold axes of symmetry (↑). There are three crystallographic axes (p.41) at 90° to each other; the parameters (p.41) on all three axes are equal.

**tetragonal system** a crystal system (↑) with one fourfold axis of symmetry (↑). There are three crystallographic axes (p.41) at 90° to each other. The parameters (p.41) on the two horizontal axes are equal but are not equal to the parameter on the vertical axis.

**hexagonal system** a crystal system (↑) with one sixfold axis of symmetry (↑). There are four crystallographic axes (p.41), three at 120° in the horizontal plane, and one vertical and perpendicular to them. The parameters (p.41) for the three horizontal axes are all equal but are not equal to the parameter for the vertical axis.

**trigonal system** a crystal system (↑) with one threefold axis of symmetry (↑). The crystallographic axes (p.41) are usually as for the hexagonal system (↑). In the past the trigonal system was included with the hexagonal system.

**rhombohedral system** = trigonal system (↑).

**orthorhombic system** a crystal system (↑) with three twofold axes of symmetry (↑). There are three crystallographic axes at 90° to each other. The parameters (p.41) are unequal.

**monoclinic system** a crystal system (↑) with one twofold axis of symmetry. There are three crystallographic axes (p.41), of which two (one of them the vertical axis) are at 90° to each other. The three parameters (p.41) are unequal.

**triclinic system** a crystal system with no axes of symmetry. There is only a centre of symmetry. There are three crystallographic axes (p.41), none of them at 90° to each other. The parameters (p.41) are all unequal.

**mineral** (*n*) a substance having a definite chemical composition (p.15), or a definite range of composition, that has been formed naturally and occurs in the Earth's crust. Most minerals have a characteristic crystal form (p.40).

**mineralogy** (*n*) the study of minerals (↑).

**specific gravity** the ratio of the mass of a substance to the mass of an equal volume of water.

**hardness** (*n*) the hardness of a mineral (↑) is measured by its ability to make a mark on the surface of another mineral. The scale of hardness that is used is due to Mohs:, ranging from 1, talc (p.61) to 10, diamond (p.48). The surfaces of minerals with a hardness of less than 6½ can be marked (scratched) with a knife. Minerals with a hardness of 2½ or less can be scratched with a finger-nail.

**Mohs' scale** *see* **hardness** (↑).

**1 talc**
**2 gypsum**
**3 calcite**
**4 fluorite**
**5 apatite**
**6 orthoclase**
**7 quartz**
**8 topaz**
**9 corundum**
**10 diamond**

**cleavage** (*n*) the cleavage of a mineral (↑) is the property of breaking along clearly marked smooth planes which are parallel to possible crystal faces (p.40). **cleave** (*v*).

**cleavage plane** a flat surface along which a mineral (↑) will cleave (↑).

**fracture** (*n*) the broken surface of a mineral. Its character can be useful in naming minerals.

**streak** (*n*) the colour of a mineral when it is in the form of a powder. This colour may be different from its colour in the mass. The streak is seen by rubbing a piece of the mineral on a rough plate called a *streak-plate*.

**crystallinity** (*n*) the degree to which a substance shows crystal form (p.40).

**crystallized** (*adj*) showing well-developed crystals (p.40).

**crystalline** (*adj*) (1) of the nature of a crystal (p.40); with regular atomic (p.152) structure; (2) composed of a mass of imperfectly formed crystal grains (p.72).

**cryptocrystalline** (*adj*) with crystals (p.40) that can be seen only under the microscope (p.147).

**amorphous** (*adj*) with no crystal (p.40) structure, as in natural glasses (p.72).

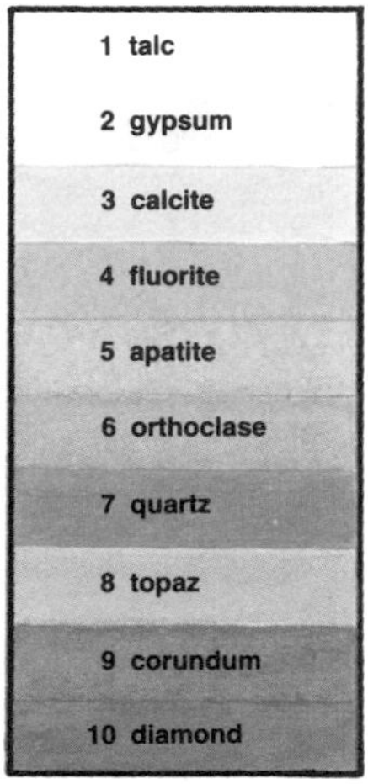

**habit** (*n*) the characteristic shapes of crystals (p.40) that are due to variations in the number, size, and shape of the crystal faces (p.40).
**equant** (*adj*) a habit (↑) in which the dimensions of a crystal (p.40) are about the same in all directions.
**tabular** (*adj*) a habit (↑) in which two dimensions of a crystal (p.40) are much greater than the third.

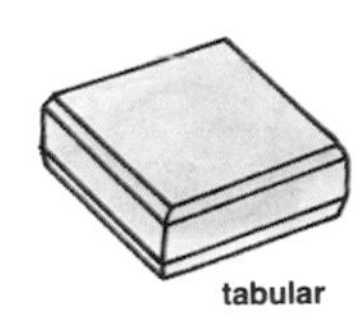
**tabular**

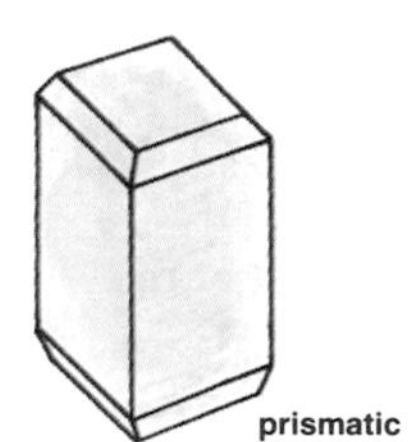
**prismatic**

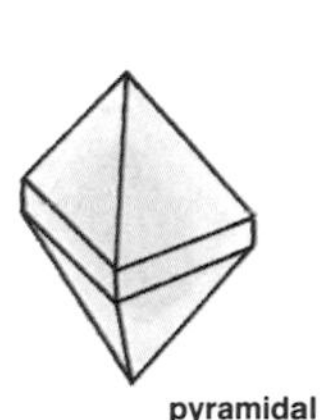
**pyramidal**

**prismatic** (*adj*) a habit (↑) in which the prism faces (p.41) of a crystal (p.40) are well shown.
**columnar** (*adj*) a habit (↑) in which the dimensions of a crystal (p.40) are almost equal in cross-section.
**pyramidal** (*adj*) a habit (↑) in which the pyramid faces (p.41) of a crystal (p.40) are well developed.
**lamellar** (*adj*) a habit (↑) in which the crystals (p.40) are in the form of thin plates or *lamellae*.
**acicular** (*adj*) a habit (↑) in which the crystals (p.40) are shaped liked needles.
**fibrous** (*adj*) a habit (↑) in which the crystals (p.40) are like threads.

**botryoidal**

---

*Terms used to describe minerals in the mass:*

**dendritic** (*adj*) shaped like a tree.
**foliated** (*adj*) in the form of thin leaves or *folia*.
**botryoidal** (*adj*) shaped like a bunch of grapes.
**euhedral** (*adj*) showing fully developed (p.40) form.
**subhedral** (*adj*) showing some signs of crystal (p.40) form.
**anhedral** (*adj*) showing no crystal (p.40) form at all.
**massive** (*adj*) not clearly crystalline (p.40).

**dendritic**

**solid solution** an ion (p.15) can take the place of another ion in a crystal lattice (p.40) if it is of about the same size and has the same charge (p.153). When this happens, the result is a solid solution. For example, the composition of the olivines (p.58) can vary continuously between $Fe_2SiO_4$ at one end of the series and $Mg_2SiO_4$ at the other end. A series of this kind is called a *solid solution series*, or *isomorphous series*.

**solid solution series** *see* **solid solution** (↑).

**isomorphous series** *see* **solid solution** (↑).

**end-member** one of the mineral compositions (p.15) at one end of a solid solution series (↑).

**zoned crystal** a crystal (p.40) of which the chemical composition (p.15) varies from the centre to the outside. There are usually zones or bands in a regular arrangement. Zoning is found where there is a solid solution series (↑).

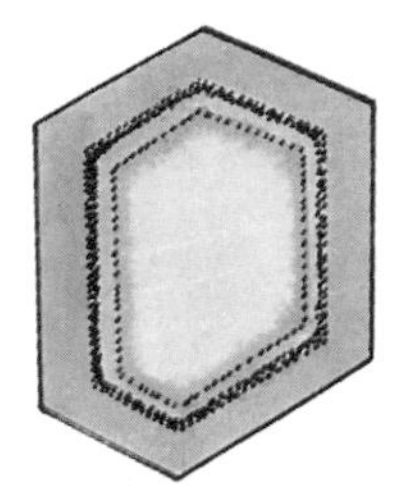

zoned crystal

**epitaxis** (*n*) the growth of one mineral upon another. **epitaxial** (*adj*).

**reaction rim** a band or zone on the outside of a crystal (p.40) formed by chemical reaction (p.17) with the material around it.

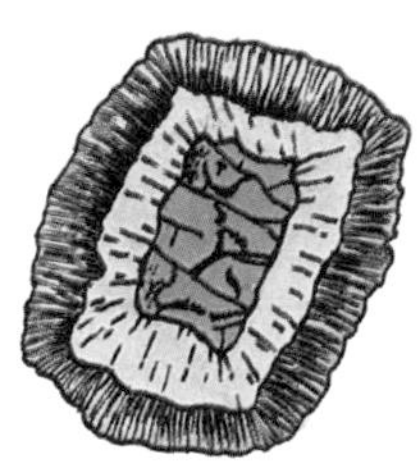

reaction rim

**inclusion** (*n*) a piece of one mineral or other substance with another mineral or rock around it on all sides, e.g. a crystal of leucite (p.58) containing augite (p.57) as inclusions.

**polymorphism** (*n*) a substance that is found in two or more forms having the same chemical composition (p.15) but with different physical properties (e.g. colour, crystal form, hardness) is said to show polymorphism; e.g. aluminium silicate, $Al_2SiO_5$, as andalusite, sillimanite, and kyanite (p.59).

**dimorphism** (*n*) a type of polymorphism (↑) in which the substance is found in two different forms; e.g. carbon as diamond and graphite (p.48); calcium carbonate as calcite and aragonite (p.51).

**pseudomorph** (*n*) a mineral in the form of another mineral. A pseudomorph may be formed by replacement; by the coating of one mineral by another; by the filling up of the space left by another mineral; or by alteration; e.g. pseudomorphs of quartz after fluorite.

**optical properties** the characters of minerals that depend upon light. Some of these can be studied only with a petrological microscope (p.147).

**isotropic** (*adj*) having the same physical properties in all directions. A mineral that is optically isotropic has the same refractive index (p.159) in all directions.

**anisotropic** (*adj*) not isotropic (↑); having physical properties that are different in different directions. A mineral that is optically anisotropic thus has a refractive index (p.159) that varies with direction in the crystal. All minerals except those belonging to the cubic system (p.43) are anisotropic.

**birefringence** (*n*) double refraction; the property of having more than one value for the refractive index (p.159); the difference between the largest and smallest values of the refractive index for a given (anisotropic (↑)) mineral. **birefringent** (*adj*).

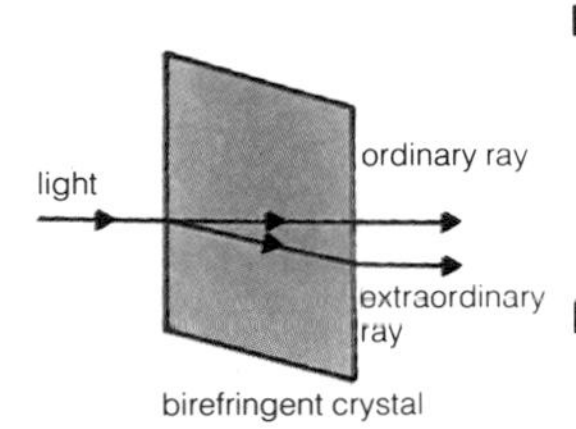


**birefringence**

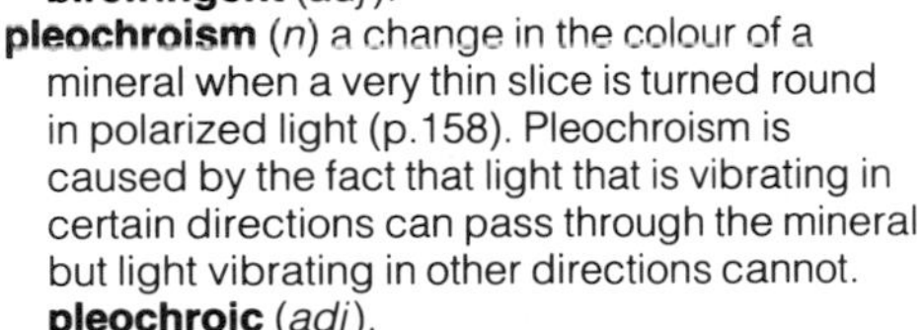

**pleochroism** (*n*) a change in the colour of a mineral when a very thin slice is turned round in polarized light (p.158). Pleochroism is caused by the fact that light that is vibrating in certain directions can pass through the mineral but light vibrating in other directions cannot. **pleochroic** (*adj*).

**dichroism** (*n*) = pleochroism (↑).

**iridescence** (*n*) a show of colours caused by optical interference (p.158) in the mineral. Iridescence is seen in calcite (p.51), mica (p.55), and other minerals. **iridescent** (*adj*).

**schiller** (*n*) a play of light that is seen on the surfaces of certain minerals when they are held at a particular angle to the light. It can resemble metallic lustre (↓). Schiller is shown by certain pyroxenes and feldspars (pp.57, 56) and by other minerals.

**lustre** (*n*) the character of the light reflected by a mineral. It may be *metallic* (like a metal); *vitreous* (like broken glass); *resinous* (like resin, the sticky material that comes out of trees); *pearly* (like pearls, the round white jewels that are found in certain shell-fish); *silky* (like silk); *adamantine* (like diamond, p.48). **lustrous** (*adj*).

**carbon minerals** the element carbon (C) is known in three different forms: graphite (↓), diamond (↓), and amorphous carbon (soot, etc.). The molecular structures (p.15) of the two minerals graphite and diamond are different and their properties differ as a result.

**graphite** (*n*) carbon with a layer structure. Very soft, with perfect cleavage; grey to black in colour. Graphite occurs in metamorphic (p.90) and igneous (p.62) rocks. *See also* **diamond** (↓).

**diamond** (*n*) pure carbon with a structure in which the atoms are connected to each other throughout the crystal lattice (p.40). Diamond is very hard and is used for jewellery and for cutting. It occurs in ultrabasic rocks (p.75) and alluvial (p.24) deposits (p.80). *See also* **graphite** (↑).

**native gold** pure gold (Au) or gold plus silver. Usually found as yellow grains in alluvial (p.24) deposits (p.80).

**native sulphur** sulphur (S); not always pure. Found at hot springs, round volcanoes, and as bedded deposits (p.80) with gypsum (p.52).

**haematite, hematite** (*n*) iron oxide, $Fe_2O_3$. It occurs as fibrous (p.45) crystals (p.40) or in masses of rounded shape, grey to black or reddish-black. An important ore (p.145) of iron.

**limonite** (*n*) a mixture of iron oxides (p.16) and hydroxides (chemical compounds (p.15) containing the OH group). Amorphous (p.44); yellow or reddish-brown to black in colour. Limonite is formed by the weathering (p.20) of minerals that contain iron. **limonitic** (*adj*).

**cassiterite** (*n*) tin oxide, $SnO_2$. Twin crystals (p.41) are common. Cassiterite occurs in acid igneous rocks (p.74), in veins (p.145), and in alluvial (p.24) deposits (p.80). An important ore (p.145) of tin.

**cuprite** (*n*) copper oxide, $Cu_2O$. Cuprite is found in the zone of weathering (p.20) of copper ores (p.145). It is itself an ore of copper.

**uraninite** (*n*) uranium oxide, $UO_2$, but not usually pure. Occurs as a primary mineral in granites (p.76) and pegmatites (p.79) and in hydrothermal veins (p.145).

**pitchblende** (*n*) a variety of uraninite (↑).

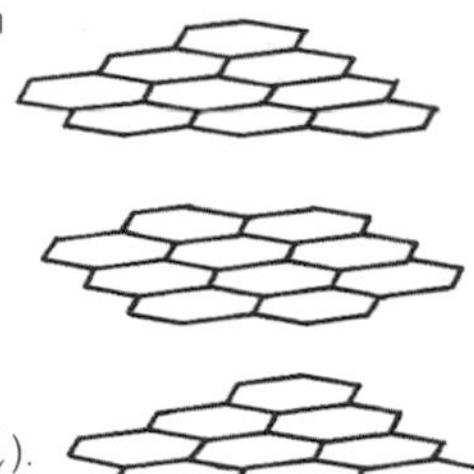

**the structure of graphite**

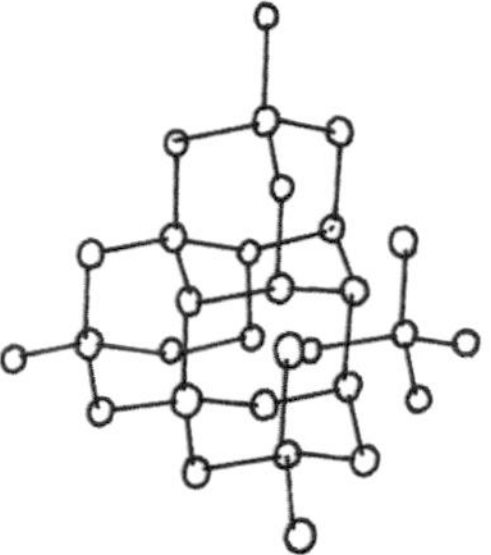

**the structure of diamond**

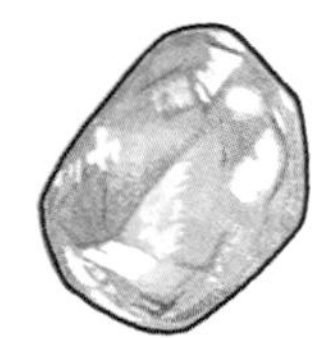

**diamond**

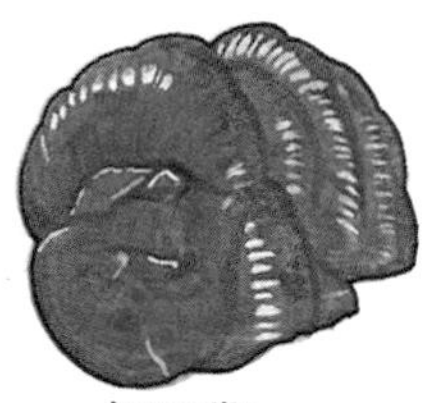

**haematite**

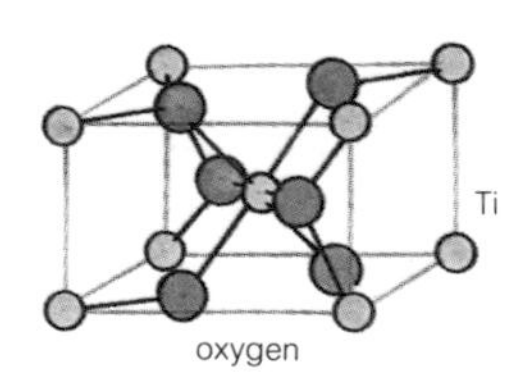

**structure of rutile, $TiO_2$**

**rutile** (*n*) titanium dioxide, $TiO_2$. Crystals are commonly acicular (p.45); twin crystals (p.41) are knee-shaped (*geniculate*). An accessory mineral (p.75) in many kinds of igneous (p.62) and metamorphic (p.90) rocks and in sediments (p.80). *See also* **anatase** (↓).

**anatase** (*n*) titanium dioxide, $TiO_2$. Occurs in metamorphic rocks (p.90) and in veins (p.145). *See also* **rutile** (↑).

**ilmenite** (*n*) an oxide of iron and titanium, $FeTiO_3$. Occurs as an accessory mineral (p.75) in basic igneous rocks (p.74), in veins, and in detrital deposits (p.85). It is the chief ore (p.145) of the metal titanium.

**corundum**

**corundum** (*n*) aluminium oxide, $Al_2O_3$. Common in barrel-shaped crystals; occurs in igneous rocks (p.62) and as a product of contact metamorphism (p.90). Very hard; used as an abrasive and for jewellery.

**spinels** (*n.pl.*) metallic oxides (p.16) occurring in basic and ultrabasic igneous rocks (pp.74, 75). Members of the spinel group include magnetite, $Fe_3O_4$, an important ore (p.145) of iron, which occurs in a variety of igneous and metamorphic rocks (p.90); spinel (in the narrower sense), $MgAl_2O_4$, found in crystalline limestones and schists (p.97); and chromite, $FeCr_2O_4$, which occurs in ultrabasic igneous rocks (p.75) and is an important ore (p.145) of chromium.

**magnetite** (*n*) *see* **spinels** (↑).

**chromite** (*n*) *see* **spinels** (↑).

**pyrolusite** (*n*) manganese dioxide, $MnO_2$; dark grey in colour. Pyrolusite is found in sediments (p.80) as nodules (p.84) – e.g. on the floor of the deep ocean – and in lakes.

**apatite** (*n*) a calcium phosphate with small amounts of fluorine, chlorine, and OH. Crystals are common. Apatite occurs in igneous rocks (p.62) as an accessory mineral; in metamorphic rocks (p.90), especially metamorphosed limestones (p.86); as veins (p.145); and as bedded deposits (p.80).

**pyromorphite** (*n*) a lead compound, $(PbCl)Pb_4(PO_4)_3$; green, yellow, or brown in colour. It occurs with lead ores (p.145).

**galena** (*n*) lead sulphide, PbS. Crystals are cubes and combinations of cubes and octahedra (p.41). Grey in colour with metallic lustre (p.47). Occurs in veins (p.145). An important ore (p.145) of lead.

**cinnabar** (*n*) mercury sulphide, HgS. Red in colour. Found in volcanic areas (p.68). An important ore (p.145) of mercury.

**pyrite** (*n*) iron sulphide, $FeS_2$. Yellow in colour with metallic lustre (p.47). Occurs in igneous rocks (p.62) as an accessory mineral (p.75); in veins (p.145); and in large deposits and in sediments formed under anaerobic conditions (p.81).

**chalcopyrite** (*n*) copper iron sulphide, $CuFeS_2$. Crystals are commonly twinned (p.41); usually massive (p.45). Yellow in colour with metallic lustre (p.47). Softer than pyrite. Occurs in veins (p.145). An important ore (p.145) of copper.

**chalcocite** (*n*) copper sulphide, $Cu_2S$. Usually massive (p.45). Crystals are commonly twinned (p.41). Occurs in veins (p.145) or beds (p.80) together with other copper minerals. An important ore (p.145) of copper.

**sphalerite, zinc blende** zinc sulphide, ZnS. Usually massive (p.45). Crystals are commonly twinned (p.41). Occurs with galena (↑) in hydrothermal and replacement deposits (p.145), in lodes and in veins (p.145). An important ore of zinc.

**zinc blende** *see* **sphalerite** (↑).

**molybdenite** (*n*) molybdenum sulphide, $MoS_2$. Usually occurs in plates or scales; very soft. Found in hydrothermal veins (p.145) and in acid igneous rocks (p.74). The chief ore of molybdenum.

**realgar** (*n*) arsenic monosulphide, AsS. Usually massive (p.45) or granular; red or orange. Occurs with orpiment (↓) in veins (p.145), in deposits of hot springs, and round volcanoes.

**orpiment** (*n*) arsenic trisulphide, $As_2S_3$. Usually foliaceous or massive (p.45); yellow in colour. Occurs in veins (p.145) and round hot springs.

**arsenopyrite** (*n*) iron arsenosulphide, FeAsS. Found in hydrothermal veins (p.145) with lead and silver; also round volcanoes.

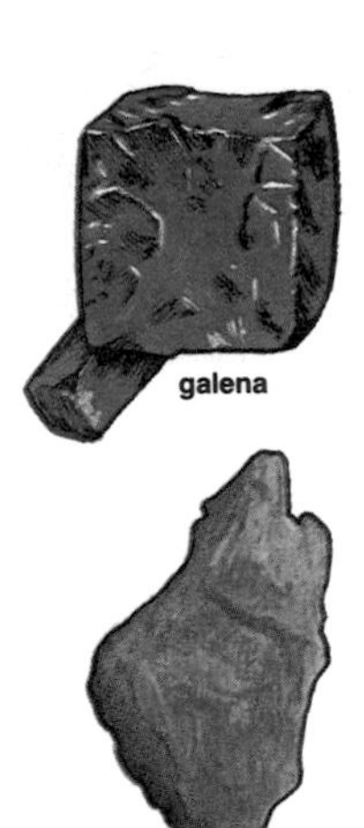


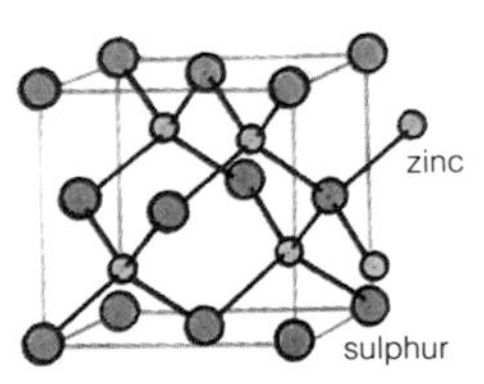

structure of sphalerite, ZnS

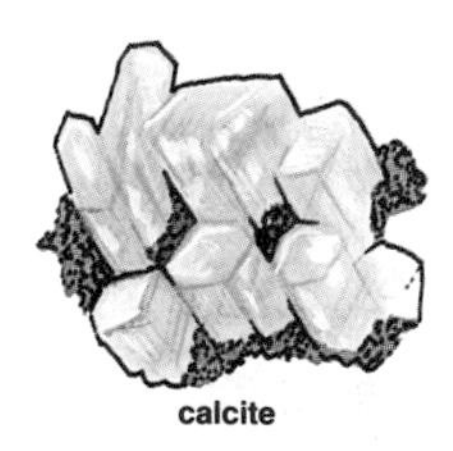
calcite

malachite

azurite

**calcite** (*n*) calcium carbonate, $CaCO_3$. Good crystals are common, with perfect cleavage (p.44). Occurs as limestone and marble; in veins (p.145); and as stalactites and stalagmites (p.21). *See also* **aragonite** (↓).

**aragonite** (*n*) calcium carbonate, $CaCO_3$. Prismatic crystals (p.41) are common, often twinned (p.41) with perfect cleavage (p.44). Less stable than calcite (↑) and is converted to calcite by heat or pressure. It occurs in sedimentary rocks (p.80).

**dolomite** (*n*) calcium magnesium carbonate, $CaMg(CO_3)_2$. Crystals show perfect cleavage (p.44). Occurs in beds formed by the alteration of limestone (p.86) and in veins (p.145).

**malachite** (*n*) hydrated basic copper carbonate, $CuCO_3.Cu(OH)_2$. Bright green in colour. Usually occurs massive (p.45). Found with copper deposits. An ore (p.145) of copper.

**azurite** (*n*) hydrated basic copper carbonate, $2CuCO_3.Cu(OH)_2$; compare with malachite (↑). Monoclinic (p.43). Occurs as crystals but is usually massive (p.45) or earthy. Deep blue in colour. Azurite occurs with other oxidized copper minerals such as malachite (↑) in the zone of weathering (p.20) of copper deposits. An ore (p.145) of copper.

**siderite** (*n*) iron carbonate, $FeCO_3$. Crystals commonly have curved faces (p.40); also massive (p.45) and oolitic (p.86). Occurs in sediments as metasomatic deposits (p.90) and as a vein mineral (p.145). Impure forms of siderite are clay-ironstone (p.88) and black-band ironstone (p.88), which are important ores (p.145) of iron.

**magnesite** (*n*) magnesium carbonate, $MgCO_3$. Crystals are rare; usually in massive (p.45) and fibrous (p.45) forms. Occurs as veins (p.145) in serpentine (p.61) and replacing dolomite (↑) and limestone (p.86).

**smithsonite** (*n*) zinc carbonate, $ZnCO_3$. Massive, botryoidal, encrusting, stalactitic, granular, or earthy (p.45); in beds and veins (p.145).

**witherite** (*n*) barium carbonate, $BaCO_3$. Crystals show repeated twinning (p.41); also massive. Occurs with galena and barite in veins (p.145).

**gypsum** (*n*) calcium sulphate, $CaSO_4.2H_2O$. Crystals are common, often twinned (p.41). Colourless and very soft. An evaporite (p.85).

**selenite** (*n*) a crystallized (p.44) variety of gypsum (↑).

**barytes, barite** (*n*) barium sulphate, $BaSO_4$. Crystals are common. Often occurs as a vein mineral (p.145).

**barite** = barytes (↑).

**anhydrite** (*n*) calcium sulphate, $CaSO_4$. Usually occurs in fibrous or lamellar form (p.45) or as granules. An evaporite; occurs in sedimentary rocks with gypsum (↑) and halite (↓).

**barytes**

**celestite, celestine** (*n*) strontium sulphate, $SrSO_4$. Usually occurs as tabular (p.45) crystals (p.40) resembling barite (↑); also massive (p.45) and fibrous (p.45). White or pale blue in colour. Celestite occurs in sedimentary rocks (p.80) and in hydrothermal veins (p.145). It is a source of strontium.

**celestine** = celestite (↑).

**wolframite** (*n*) tungstate of iron and manganese, $(Fe,Mn)WO_4$. Occurs as tabular crystals (p.45) or prismatic crystals (p.45); also massive (p.45). Wolframite is found in pneumatolytic (p.63) veins and hydrothermal veins (p.145).

**scheelite** (*n*) calcium tungstate, $CaWO_4$. Usually massive or granular (p.45). Found in pneumatolytic and hydrothermal deposits (p.145).

**fluorite, fluorspar** (*n*) calcium fluoride, $CaF_2$. Crystals are common, usually cubes. Occurs in veins (p.145) and replacement deposits (p.145) and in pegmatites (p.79).

**fluorspar** (*n*) = fluorite (↑).

**fluorite**

**halite, rock salt** (*n*) sodium chloride (common salt), NaCl. Crystals are common. Colourless or white when pure. Halite occurs as beds which are produced by the evaporation (p.156), of sea water. It can flow under high pressure; salt domes (p.131) then result which in their form resemble igneous intrusions (p.64).

**rock salt** = halite (↑).

**sylvite** (*n*) potassium chloride, KCl. It occurs with halite (↑), etc.

**silicate structures** most silicate minerals are built up from $SiO_4$ tetrahedra (p.41) in which four oxygen atoms (p.152) are arranged round a silicon atom. These tetrahedra can be joined together to form larger groups: rings, chains (single and double), layers (sheets), and frameworks in three dimensions. It is also possible for some of the silicon atoms in the tetrahedra to be replaced by aluminium atoms, which have a similar ionic radius (p.15). Various metals can be fitted into the octahedral (p.41) spaces between the $SiO_4$ tetrahedra, e.g. magnesium, titanium, iron, aluminium, sodium, and potassium. (Aluminium can thus be included in the structure in two different ways.) A great many structures and compositions (p.15) are therefore possible and the actual compositions of silicate minerals can be highly complicated.

**nesosilicate** $SiO_4^{4-}$

**sorosilicate** $SiO_7^{6-}$

**nesosilicates** (*n.pl.*) silicates (p.16) containing separate $SiO_4$ ions. These are the simplest silicate structures. The olivines, zircon, kyanite, sillimanite, and andalusite are all nesosilicates. The structure of olivine (p.58) can, for example, be thought of as a collection of $SiO_4$ groups with $Mg^{2+}$ (magnesium) ions in the holes between them, each $Mg^{2+}$ ion being surrounded by six oxygen atoms.

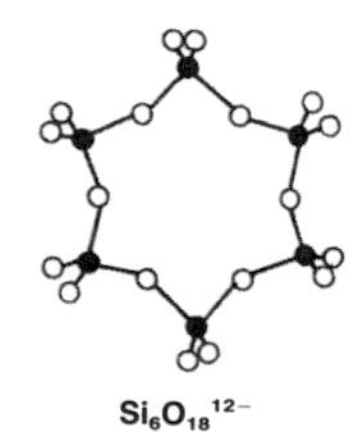

$Si_6O_{18}^{12-}$

**cyclosilicate**
ring structure

$Si_3O_9^{6-}$

**sorosilicates** (*n.pl.*) silicates (p.16) in which two $SiO_4$ tetrahedra (p.41) share one oxygen atom between them. The unit is thus $Si_2O_7$. Melilite, $Ca_2MgSi_2O_7$, is an example.

**cyclosilicates** (*n.pl.*) silicates with ring structures in which two oxygen atoms of each $SiO_4$ tetrahedron are shared with other $SiO_4$ tetrahedra. The general formula is thus $(SiO_3)^{2n-}$. The rings can have three, four, or six $SiO_4{}^n$ units. The structure with six $SiO_4$ units occurs in beryl, $Be_3Al_2Si_6O_{18}$, and cordierite (p.60). Ring structures are not, however, common.

**ring silicates** = cyclosilicates (↑).

**inosilicates** (*n.pl.*) silicates with a single chain structure. Two oxygen atoms of each $SiO_4$ unit are shared with other $SiO_4$ tetrahedra (p.41). Single or simple chains have the general formula $(SiO_3)_n$. The pyroxenes (p.57) are of this type. Double chains are also possible. These have the general formula $(Si_4O_{11})_n$. The amphiboles (p.57) are of this type. The two types of chain structure account for the different cleavages (p.44) of the pyroxenes and amphiboles.

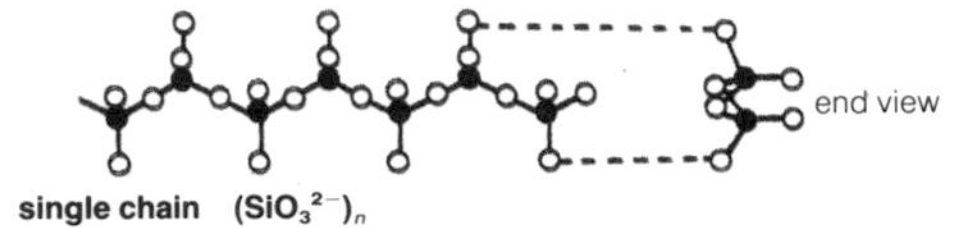


**single chain** $(SiO_3^{2-})_n$

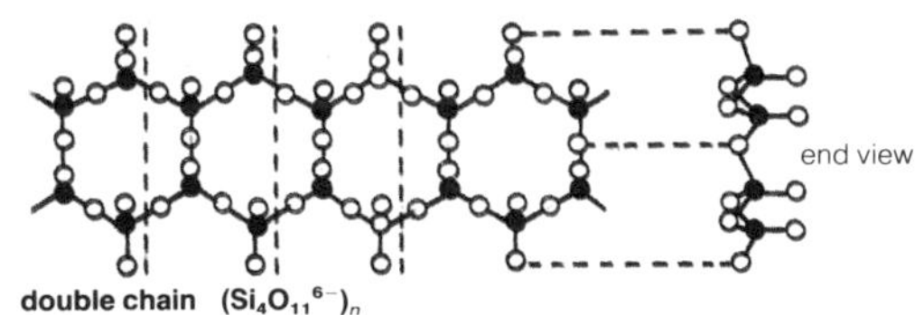


**double chain** $(Si_4O_{11}^{6-})_n$

**phyllosilicates** (*n.pl.*) in these silicate structures (p.53) layers or sheets are built up of $SiO_4$ tetrahedra (p.41) in which three oxygen atoms of each $SiO_4$ unit are shared with those next to it. The most important type of layer structure in mineralogy is one in which one or two layers of this kind are combined with layers of hydroxyl (OH) groups and with magnesium or aluminium (Mg or Al) atoms. The micas (↓) are of this type. Talc (p.61), the clay minerals (p.61), and chlorite (p.61) also have layer structures.

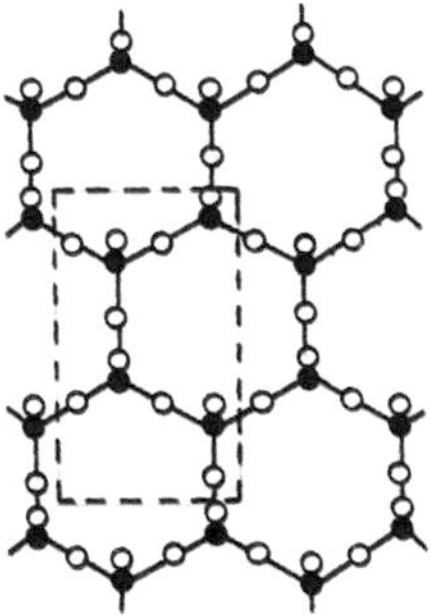

**phyllosilicates** $(Si_4O_{10}^{4-})_n$

**sheet silicates** = phyllosilicates (↑).

**layer lattice silicates** = phyllosilicates (↑).

**tectosilicates** (*n.pl.*) silicate structures (p.53) in which there is a framework or network of silicon and oxygen atoms. Some of the silicon atoms in minerals are replaced by aluminium atoms. Positive ions (p.15) such as $Na^+$ and $Ca^{2+}$ are then also present. The most important minerals with framework structures are the feldspars (p.56).

**quartz**

**quartz** (*n*) a hard, glass-like mineral; chemical composition: silica, $SiO_2$. The structure of quartz is a framework of $SiO_4$ tetrahedra (p.41).

**alpha quartz** or **low quartz** found in a great variety of igneous (p.62) and metamorphic rocks (p.90), as veins (p.145), and in sandstones (p.87).

**beta quartz** or **high quartz** formed at 573°C.

**tridymite** a form of quartz (↑) stable above 870°C.

**cristobalite** a form of quartz (↑) stable above 1470°C.

**coesite** a polymorph (p.46) of silica formed at very high pressures.

**chalcedony** a cryptocrystalline (p.44) variety of silica.

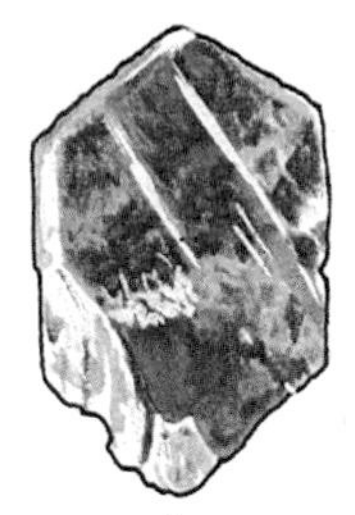

**mica**
muscovite

**micas** (*n.pl.*) a group of rock-forming minerals made up of layers of $SiO_4$ tetrahedra (p.41); chemically they are silicates of aluminium and potassium with hydroxyl (OH) groups; magnesium and iron are present in the dark micas. Most of the micas are monoclinic (p.43); all show perfect cleavage (p.44).

**muscovite** (*n*) 'white mica': $KAl_2(AlSi_3)O_{10}(OH)_2$. Colourless or light in colour. Found in acid igneous rocks (p.74), metamorphic rocks (p.90), and sedimentary rocks (p.80).

**biotite** (*n*) brown or black mica (↑) containing iron and magnesium: $K(Mg,Fe^{2+})_3(Al,Fe^{3+})Si_3O_{10}(OH)_2$. Common in igneous (p.62) and metamorphic rocks (p.90).

**phlogopite** (*n*) a brown or black mica (↑) containing iron and magnesium: $K(Mg,Fe^{2+})_3(AlSi_3)O_{10}(F,OH_2)$. It is common in igneous rocks (p.62).

**lepidolite** (*n*) a light mica (↑) containing lithium, $K(Mg,Li,Al)_3(Al,Si_3)O_{10}(OH,F)_2$. Violet in colour. Lepidolite occurs in pegmatites (p.79).

**glauconite** (*n*) a green mineral closely related to the micas (↑). It occurs in marine sediments (p.80).

**ferromagnesian minerals** a general term for silicate minerals that contain more than small amounts of iron or magnesium, or both. The ferromagnesian minerals include olivine (p.58), augite (p.57), hornblende (p.57), and biotite (↑).

**feldspars, felspars** (*n.pl.*) silicates (p.16) of sodium, potassium, calcium, and barium. Their basic structure is a network of $SiO_4$ tetrahedra (p.41). All feldspars are triclinic or monoclinic (p.43). They occur in rocks of all types.

**alkali feldspars** feldspars containing sodium silicate, $NaAlSi_3O_8$ ('soda feldspar') or potassium silicate, $KAlSi_3O_8$ ('potash feldspar'), or the two together. The group includes orthoclase (↓), adularia (↓), sanidine (↓), and microcline (↓).

**orthoclase** (*n*) an alkali feldspar (↑) $KAlSi_3O_8$. Monoclinic (p.43). Twin crystals (p.41) are common. A characteristic mineral of granites (p.76).

**adularia** (*n*) a low-temperature form of $KAlSi_3O_8$ found in veins (p.145), etc.

**sanidine** (*n*) a high-temperature form of $KAlSi_3O_8$ found in lavas (p.70) and dykes (p.67). On slow cooling it changes to orthoclase (↑).

**microcline** (*n*) the triclinic (p.43) form of $KAlSi_3O_8$; sodium is usually present as well. Microcline shows characteristic complex twinning (p.41).

**potash feldspar** potassium feldspar, $KAlSi_3O_8$; an alkali feldspar (↑).

**soda feldspar** sodium feldspar, $NaAlSi_3O_8$; an alkali feldspar (↑).

**plagioclase feldspar** a series of triclinic (p.43) feldspars (↑) with albite, $NaAlSi_3O_8$, and anorthite, $CaAl_2Si_2O_8$, as the end-members (p.46). Lamellar twinning (p.41) is common. Plagioclase is found in igneous rocks (p.62) and metamorphic rocks (p.90).

**albite** (*n*) *see* **plagioclase feldspar** (↑).

**oligoclase** (*n*) plagioclase feldspar (↑) with 70–90% albite (↑).

**andesine** (*n*) plagioclase feldspar (↑) with 50–70% albite (↑).

**labradorite** (*n*) plagioclase feldspar (↑) with 30–50% albite (↑).

**bytownite** (*n*) plagioclase feldspar (↑) with 10–30% albite (↑).

**anorthite** (*n*) *see* **plagioclase feldspar** (↑).

**perthite** (*n*) an intergrowth (p.73) of two feldspars. **perthitic** (*adj*).

**feldspar**
microcline feldspar

**pyroxenes** (*n.pl.*) inosilicates (p.54) with a single-chain structure of $SiO_4$ tetrahedra (pp.41, 53). Most pyroxenes are monoclinic (p.43). All have cleavage (p.44) at 90°, unlike the amphiboles (↓), which have a cleavage at 124°. Pyroxenes are found in basic and ultrabasic igneous rocks (p.75) and in metamorphic rocks (p.90).

**clinopyroxenes** (*n.pl.*) a general name for monoclinic (p.43) pyroxenes (↑). The most common of these is augite, $(Ca,Mg,Fe)(Si,Al)_2O_6$. Others include diopside, $CaMgSi_2O_6$, and the alkali-pyroxenes, which contain Na, Fe, Al, or Li.

**orthopyroxenes** (*n.pl.*) a general name for orthorhombic (p.43) pyroxenes (↑). They form a solid-solution series (p.46) with enstatite, $MgSiO_3$, and ferrosilite, $FeSiO_3$, as the end-members (p.46).

**augite** (*n*) *see* **clinopyroxenes** (↑).

**diopside** (*n*) *see* **clinopyroxenes** (↑).

**alkali pyroxenes** *see* **clinopyroxenes** (↑).

**enstatite** (*n*) *see* **orthopyroxenes** (↑).

**ferrosilite** (*n*) *see* **orthopyroxenes** (↑).

**amphiboles** (*n.pl.*) inosilicates (p.54) with a double-chain structure of $SiO_4$ tetrahedra (p.41). Most amphiboles are monoclinic (p.43); a few are orthorhombic (p.43). All have a cleavage (p.44) at 124°, unlike the pyroxenes (↑), which have a cleavage at 90°. Amphiboles are found in plutonic (p.64) and other igneous rocks (p.62), in metamorphic rocks (p.90), and in sediments (p.80). Fibrous (p.45) forms of amphibole are included in the asbestos group of minerals (p.61). The monoclinic amphiboles include *hornblende*, an important rock-forming mineral, $(Ca,Na)_{2-3}(Mg,Fe^{2+},Fe^{3+}Al)_5(Al,Si)_8O_{22}(OH)_2$, *tremolite* and *actinolite*, which are found in metamorphic rocks, and *glaucophane* and *riebeckite*, which are found in igneous rocks.

augite

**pyriboles** (*n,pl.*) a general term for pyroxenes (↑) and amphiboles (↑) together.

**hornblende** (*n*) *see* **amphiboles** (↑).

**tremolite** (*n*) *see* **amphiboles** (↑).

**actinolite** (*n*) *see* **amphiboles** (↑).

**glaucophane** (*n*) *see* **amphiboles** (↑).

**riebeckite** (*n*) *see* **amphiboles** (↑).

hornblende

**olivines** (*n.pl.*) a group of magnesium and iron silicate minerals built up of isolated $SiO_4$ tetrahedra (pp.41, 53) joined by cations (p.15). All olivines are orthorhombic (p.43). Crystals are green or green brown with a glassy lustre (p.47) and smoothly curved fracture (p.44). The olivines form a solid-solution series (p.46) with forsterite, $Mg_2SiO_4$, and fayalite, $Fe_2SiO_4$, as the end-members (p.46). Olivines are important minerals in basic and ultrabasic igneous rocks (pp.74, 75).

**forsterite** (*n*) *see* **olivines** (↑).

**fayalite** (*n*) *see* **olivines** (↑).

**garnets** (*n.pl.*) a group of silicate minerals with the general formula $R_3^{2+}R_2^{3+}(SiO_4)_3$, where $R^{2+}$ is Ca, Mg, $Fe^{2+}$, or Mn and $R^{3+}$ is Al, Cr, or $Fe^{3+}$. In the garnets $SiO_4$ tetrahedra (p.41) are packed together with the $R^{2+}$ and $R^{3+}$ ions between them. Garnets are generally cubic (p.43). They have high refractive indices (p.159) and are strongly coloured (red, green, or black). They have no cleavage (p.44). Garnets occur in igneous (p.62) and metamorphic rocks (p.90), e.g. in schists (p.97).

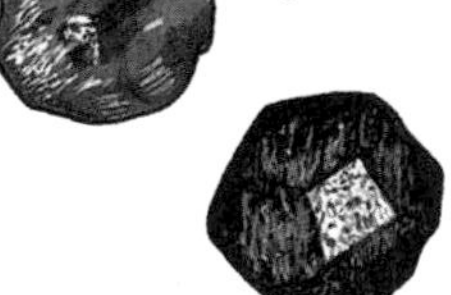

garnets

**pyrope** (*n*) a red garnet (↑), $Mg_3Al_2(SiO_4)_3$.

**almandine** (*n*) a red garnet (↑), $Fe_3Al_2(SiO_4)_3$.

**spessartite** (*n*) a red garnet (↑), $Mn_3Al_2(SiO_4)_3$.

**grossularite** (*n*) a green garnet (↑), $Ca_3Al_2(SiO_4)_3$.

**andradite** (*n*) a red or green garnet (↑), $Ca_3Fe_2(SiO_4)_3$.

**uvarovite** (*n*) a green garnet (↑), $Ca_3Cr_2(SiO_4)_3$.

**feldspathoids** (*n.pl.*) sodium and potassium silicate minerals (p.44) with framework structures (p.122). The feldspathoids are closely related to the feldspars (p.56) but are chemically undersaturated (p.74), i.e. they are never present at the same time as quartz (p.55). Most feldspathoids are cubic or hexagonal (p.43).

**nepheline** (*n*) a feldspathoid (↑), $NaAlSiO_4$.

**leucite** (*n*) a feldspathoid (↑), $KAlSi_2O_6$.

**sodalite** (*n*) a complex feldspathoid (↑).

**nosean** (*n*) a complex feldspathoid (↑).

**haüyne** (*n*) a complex feldspathoid (↑).

**cancrinite** (*n*) a complex feldspathoid (↑).

**aluminium silicate minerals** there are three common minerals with the chemical composition $Al_2SiO_5$: andalusite, sillimanite, and kyanite (↓). They are found in metamorphic rocks (p.90).

**andalusite** (*n*) aluminium silicate, $Al_2SiO_5$. Orthorhombic (p.43); crystals are prismatic (p.45). Andalusite is formed at high temperature and low stress (p.122) and is typical of thermal metamorphism (p.90), especially of argillaceous rocks (p.85). *See also* **sillimanite** (↓), **kyanite** (↓).

andalusite

**sillimanite** (*n*) aluminium silicate, $Al_2SiO_5$. Orthorhombic (p.43). Commonly occurs as fibrous masses (p.45). Sillimanite is formed at high temperatures and is found in regionally metamorphosed rocks of high grade (pp.90, 91). It gives its name to the sillimanite zone (p.91). *See also* **andalusite** (↑), **kyanite** (↓).

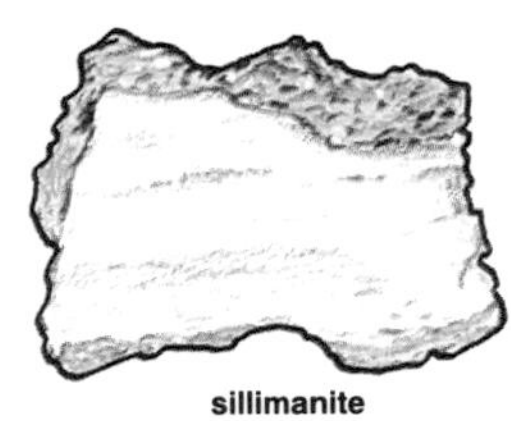

sillimanite

**kyanite** (*n*) aluminium silicate, $Al_2SiO_5$. Triclinic (p.43). Commonly occurs as blue crystals. Kyanite does not change under stress and is characteristic of metamorphic rocks of intermediate grade (pp.90, 91). It gives its name to the kyanite zone (p.91). *See also* **andalusite** (↑), **sillimanite** (↑).

**staurolite** (*n*) a complex iron aluminium silicate. Monoclinic (p.43), though it appears to be orthorhombic. Staurolite is found in regionally metamorphosed argillaceous rocks (p.85) of medium grade (p.91). It gives its name to the staurolite zone (p.91).

**sphene** (*n*) a calcium titanium silicate, $CaTiSiO_5$. Twin crystals (p.41) are common; it also occurs in massive form (p.45). Sphene occurs as an accessory mineral (p.75) in acid igneous rocks (p.74) and in contact-metamorphosed limestones (pp.86, 90).

**titanite** (*n*) = sphene (↑).

**zircon** (*n*) zirconium silicate, $ZrSiO_4$. Occurs as prismatic crystals (p.45). Zircon is found in acid igneous rocks (p.74), in metamorphic rocks (p.90), and in sediments (p.80).

**topaz** (*n*) an aluminium silicate, $Al_2SiO_4(OH,F)_2$. It is found in acid igneous rocks (p.74), in tin veins (p.145), and in contact zones (p.92).

**cordierite** (*n*) an aluminium iron magnesium silicate. The structure is built up of $SiO_4$ units in a six-membered ring (p.53). Orthorhombic (p.43); usually granular or massive (p.45); blue in colour and glassy. Cordierite is found in rocks that have been contact metamorphosed (p.90) and in thermally metamorphosed argillaceous rocks (p.85).

**beryl** (*n*) a beryllium aluminium silicate, $Be_3Al_2Si_6O_{18}$. The structure is a six-membered ring (p.53). Hexagonal (p.43); crystals are common. Very hard; used for jewellery. Beryl occurs as an accessory mineral in acid igneous rocks (p.74) and in metamorphic rocks (p.90).

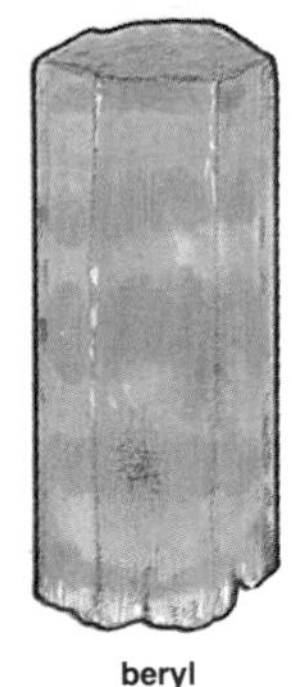

beryl

**tourmaline** (*n*) a complex borosilicate containing sodium, magnesium or iron, manganese, and lithium. Its structure is of the six-membered ring type (p.53). Hexagonal (p.43); needle-shaped crystals are common, often in groups. Tourmaline occurs as an accessory mineral (p.75) in igneous (p.62) and metamorphic rocks (p.90) and in sediments (p.80).

**idocrase** (*n*) a complex calcium aluminium magnesium silicate (p.16) formed by the contact metamorphism (p.90) of impure limestones (p.86).

**vesuvianite** (*n*) = idocrase (↑).

**epidotes** (*n.pl.*) silicate minerals with $SiO_4$ and $Si_2O_7$ units. Most epidotes contain calcium and/ or aluminium, iron, and manganese. Some are orthorhombic, others are monoclinic (p.43). The epidotes include zoisite, a calcium aluminium silicate, and epidote (in the narrower sense), a calcium iron aluminium silicate. Zoisite is orthorhombic (p.43) but there is also a monoclinic form, *clinozoisite*. Epidote is monoclinic (p.43). Epidotes are found in metamorphic rocks (p.90) of lower grade (p.91), in igneous rocks (p.62), and in sandstones (p.87).

**piedmontite** (*n*) an epidote (↑) containing manganese.

**allanite** (*n*) an epidote (↑) containing cerium.

**orthite** (*n*) = allanite (↑).

**zoisite** (*n*) *see* **epidotes** (↑).

**clinozoisite** (*n*) *see* **epidotes** (↑).

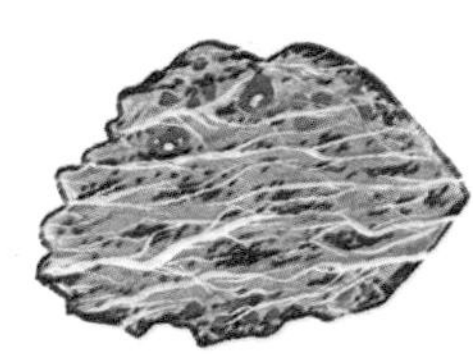
serpentine

asbestos

**serpentine** (*n*) a hydrous (p.157) magnesium silicate, green or nearly black, usually with fine lines or spotted. It occurs in metamorphosed (p.90) basic and ultrabasic igneous rocks (pp.74–5) by the alteration of olivines (p.58) and pyroxenes (p.57).

**talc** (*n*) a hydrated magnesium silicate. It is very soft and is usually massive (p.45). Talc occurs in metamorphosed (p.90) basic rocks (p.74) with serpentine (↑).

**asbestos** (*n*) (1) the fibrous (p.45) forms of various silicate minerals, including certain amphiboles (p.57) and serpentine (↑); (2) (in mineralogy) the fibrous (p.45) forms of amphiboles only.

**clay minerals** a group of silicate minerals (p.16) with sheet or layer structures (p.54) formed by the weathering (p.20) of other silicate minerals. Clay minerals occur as very fine particles and can readily take up water.

**kaolinite** (*n*) a group of clay minerals (↑) with the formula $Al_4Si_4O_{10}(OH)_8$. They are formed by the decomposition of feldspars (p.56).

**illite** (*n*) a group of clay minerals (↑) formed by the decomposition of certain silicate minerals.

**montmorillonite** (*n*) a group of clay minerals formed by the alteration of certain silicate minerals.

**smectite** = montmorillonite (↑).

**fullers' earth** a fine earthy material containing montmorillonite (↑).

**vermiculite** (*n*) a group of clay minerals related to chlorite and montmorillonite (↑). Vermiculite is formed by the alteration of micas, especially biotite, and other minerals.

**chlorite** (*n*) a group of minerals with a phyllosilicate structure (p.54) related to talc (↑). They occur in igneous rocks (p.62), where they are formed by alteration of ferromagnesian minerals (p.55), and in metamorphic rocks (p.90).

**zeolites** (*n*) a group of hydrous (p.157) aluminium silicate minerals containing sodium, potassium, calcium, and barium. These positive ions (p.15) can easily be replaced. Zeolites occur in volcanic rocks (p.68) and in hydrothermal veins (p.145).

**rock** (*n*) in geology, any natural material formed of minerals (p.44), or less commonly of a single mineral, whether solid or not. Rocks are divided into three main classes: igneous (↓), sedimentary (p.80), and metamorphic (p.90).
**petrology** (*n*) the study of rocks: their origin, their occurrence, and what they are made of. **petrological** (*adj*).
**petrography** (*n*) the description of rocks and their grouping into classes. A branch of petrology (↑). **petrographic** (*adj*).
**igneous** (*adj*) igneous rocks are, with certain exceptions, the crystalline (p.40) or glassy rocks that have solidified from magma (↓): the lavas that have been poured out at the surface of the Earth and the rocks that have solidified at various depths in the crust (p.9).
**petrogenesis** (*n*) the origin of rocks and the ways in which they are formed. The word is generally used only in igneous petrology (↑).
**magma** (*n*) rock material in a molten state. Magmas consist of silicates (p.16), water, and gases at high temperatures. **magmatic** (*adj*).
**magmatism** (*n*) the formation and movement of magma (↑) and of igneous rocks (↑) in the Earth's crust (p.9).
**magmatic differentiation** the separation of a magma (↑) into two or more parts, called *fractions*, with differing compositions (p.15). The separate fractions may then crystallize out (p.40) as rocks of different chemical composition (p.15) containing different minerals (p.44).
**fraction** (*n*) *see* magmatic differentiation (↑).
**hybrid** (*n, adj*) a hybrid rock is an igneous rock (↑) that has been produced by the mixing together of two magmas (↑) of different compositions or by the assimilation (↓) of other rocks by a magma. **hybridization** (*n*).
**enrichment** (*n*) a process in which the amount of one mineral or chemical element in a rock is increased in relation to the rest.
**assimilation** (*n*) the process of taking material (usually solid rock) into an igneous rock (↑) by melting it.

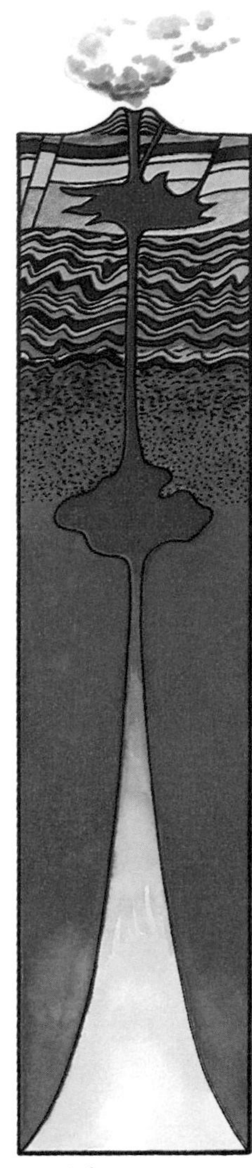

**ascent of magma**

**pneumatolysis** (*n*) changes produced by hot gases (e.g. fluorine, hydrofluoric acid) given off by a magma in the later stages of cooling and solidification. **pneumatolytic** (*adj*).

**hydrothermal** (*adj*) caused by the action of water at high temperatures during the formation of igneous rocks. *See also p.145.*

**diffusion** (*n*) the spreading out of ions (p.15), atoms (p.152), or molecules (p.15) into a liquid, a gas, or a porous (p.84) solid (such as a rock). The process of diffusion tends to distribute a substance more evenly throughout the system. **diffuse** (*v*), **diffused** (*adj*).

**granitization** (*n*) a process by which solid sedimentary rocks (↑) that show bands changed into granite (p.76) by the action of liquids or gases without first being turned into magma (↑).

**layering, banding** (*n*) terms used to describe bodies of igneous rock (p.62) that show bands when seen in a vertical section (p.147). 'Layering' is a better term than 'banding'. Layering may be the result of gravity separation (↓) or of other physical processes.

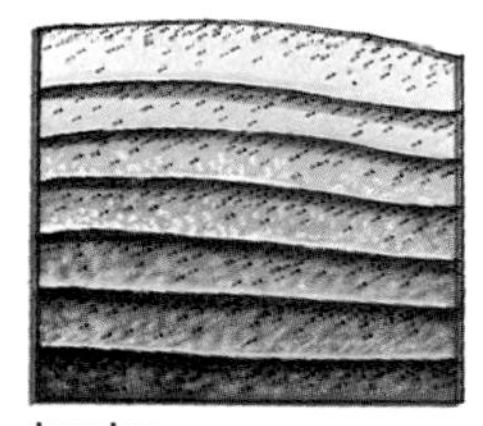

**layering**

**banding** (*n*) = layering (↑).

**gravity separation** it has generally been thought that if the crystals (p.40) that are first to form in a magma (↑) are heavier or lighter than the liquid they will sink or rise through it and will gather together at a lower or higher level in the magma chamber (p.64). This process is called *gravity separation*. If the movement of the crystals is downward the term 'crystal settling' can be used.

**crystal settling** *see* **gravity separation** (↑).

**cumulate** (*n*) an igneous rock that is thought to have formed by crystal settling (↑).

**compositional zoning** differences in the temperature and pressure in different parts of a magma chamber (p.64) can result in variation in the composition of the magma. The magma erupted will then vary in composition with time.

**devitrification** (*n*) the formation of crystals (p.40) in a glass (p.77), which in time may become completely crystalline. **devitrified** (*adj*).

**igneous intrusion** a body of igneous rock (p.62) that has been put into place among rocks that were already there before it crystallized (p.44). **intrude** (*v*), **intrusive** (*adj*), **intrude** (*v*).

**emplacement** (*n*) any process by which an igneous rock is put into place. The word 'emplacement' can be used without suggesting any particular method of emplacement. **emplace** (*v*).

**magma chamber** a space below the Earth's surface containing magma (p.62).

**injection** (*n*) the intrusion (↑) of magma into rocks that are already in place. **inject** (*v*).

**hypabyssal** (*adj*) at no great depth. The word is used to describe bodies of igneous rock, such as dykes (p.67) and sills (p.66), that have formed at greater depths than volcanic rocks (p.68) but not at such great depth as plutonic rocks (↓).

**plutonic rocks** those igneous rocks (p.62) that have formed from magma (p.62) or by chemical alteration – metasomatism (p.90) – at great depth in the Earth's crust.

**pluton** (*n*) a body of plutonic rock (↑). The term 'pluton' is now often used to mean a granitic (p.76) rock body of roughly circular shape in plan that has been emplaced (↑) at a relatively low temperature.

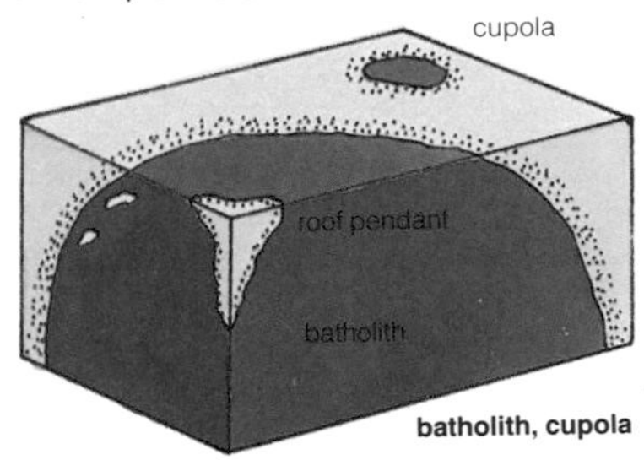


**batholith, cupola**

**batholith, bathylith** (*n*) a pluton (↑) with a large outcrop (p.122) and no visible base. Usually a granite (p.76). **batholithic** (*adj*).

**cupola** (*n*) an igneous (p.62) rock-body, round in plan, which is joined below to a larger body of igneous rock, such as a batholith (↑).

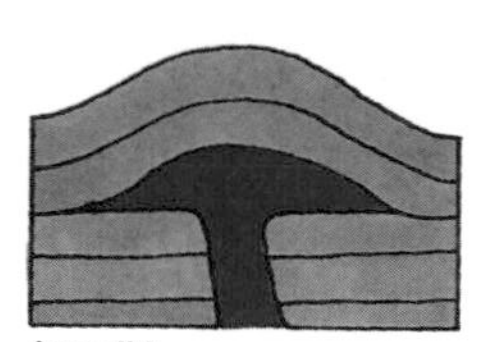
laccolith

**laccolith** (*n*) an igneous intrusion (↑) with a flat floor and a rounded roof which has pushed up the sediments (p.80) above it into the shape of a dome.

**stock** (*n*) an igneous intrusion (↑) like a batholith (↑) but smaller (less than 100 $km^2$ in area) and with a more or less circular shape in plan.

**boss** (*n*) an igneous intrusion (↑) that is circular in plan and has steeply dipping contacts (p.148) with the country-rock (↓).

**multiple intrusion** an igneous intrusion (↑) that has been put into place in more than one injection (↑) of magma (p.62). The material is usually all of the same kind. Multiple intrusions are usually sills or dykes (pp.66, 67).

**country-rock** (*n*) the rock or rocks into which an igneous intrusion (↑) or a mineral vein (p.145) is emplaced.

**xenolith** (*n*) a piece of 'foreign' rock that is enclosed by an igneous rock body, e.g. a piece of country-rock broken off from the wall of the intrusion. **xenolithic** (*adj*).

stoping

**stoping, magmatic stoping** (*n*) a way in which an igneous rock (p.62) may be emplaced (↑) by forcing its way into joints (p.21) in the country-rock (↑) and pushing out blocks of it, which then sink into the magma (p.62) to be assimilated (p.62).

**subjacent** (*adj*) without a known floor. The word is used to describe large igneous intrusions (↑) for which there is no sign of a base.

**cross-cutting** (*adj*) cutting across other, earlier rocks.

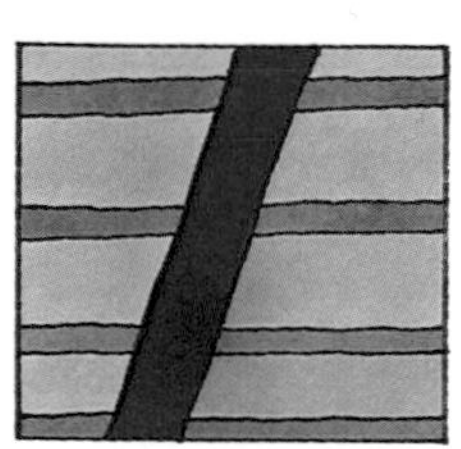
cross-cutting

**discordant** (*adj*) a word used to describe an igneous rock that cuts across the bedding (p.80) or the foliation (p.95) of the rocks into which it is intruded. *See also* **concordant** (↓).

**concordant** (*adj*) a word applied to an igneous rock that is parallel to the bedding (p.80) or the foliation (p.95) of the rocks into which it is intruded. *See also* **discordant** (↑).

**schlieren** (*n.pl.German*) Long-drawn-out lines or areas in an igneous rock. Schlieren are usually of different composition (p.15) from the rest of the rock.

**apophysis** (*n*) (*apophyses*) a vein (p.145) or branch of an igneous rock (p.62) that is joined to a larger body.
**tongue** (*n*) = apophysis (↑).
**lit-par-lit** (*French*) 'Bed by bed'. *Lit-par-lit* intrusion (p.64) is the injection (p.64) of magma (p.62) along bedding-planes (p.80) or foliation-planes (p.95) to give a rock with thin, closely spaced layers of igneous material, e.g. a foliated gneiss (p.97).
**chilled margin, chilled zone** a border of fine-grained rock at the edge of an igneous intrusion (p.64). The fine grain is the result of rapid cooling of the magma (p.62) by the country-rock (p.65).

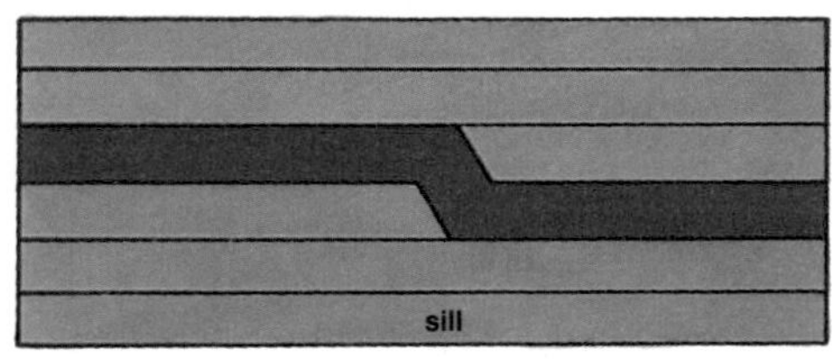
sill

**sill** (*n*) an igneous intrusion (p.64) in the form of a concordant (p.65) sheet; usually more or less horizontal.
**lopolith** (*n*) a large igneous intrusion (p.64) shaped like a nearly flat dish. A lopolith is generally concordant (p.65) and its width is about ten to twenty times its thickness.
**phacolith** (*n*) an igneous intrusion (p.64) in folded sedimentary rocks (p.80) that is convex upwards and concave below.

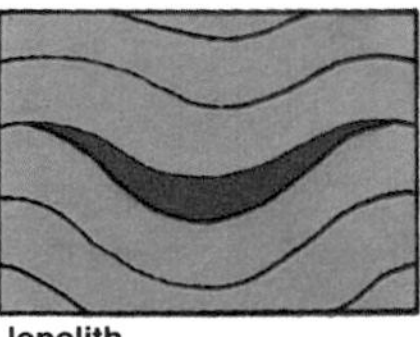
lopolith

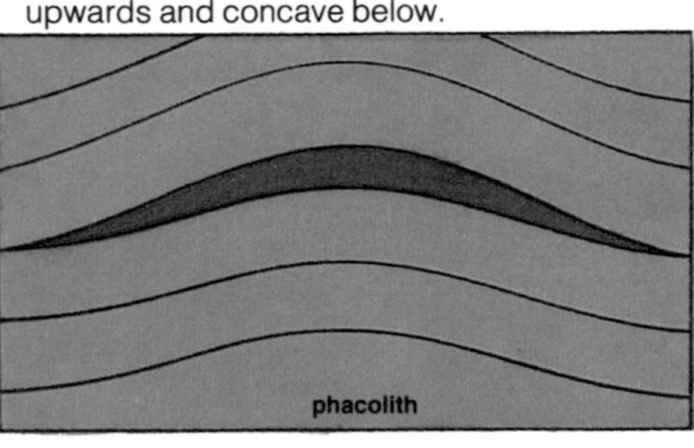
phacolith

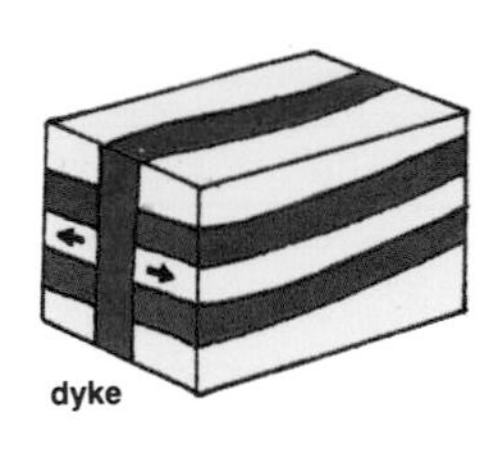
dyke

**dyke** (*n*) an igneous intrusion (p.64) like a wall that is discordant (p.65) to the country-rock (p.65).
**dike** (*n*) American spelling of dyke (↑).
**dyke-swarm** (*n*) a group of dykes (↑) injected at about the same time and either parallel to each other or in lines meeting at a point.
**ring-dyke** (*n*) a dyke (↑) with an outcrop (p.122) in the shape of a curve or a circle.
**ring-complex** (*n*) a group of igneous intrusions (p.64) occurring together that have an outcrop (p.122) shaped like a ring, e.g. a complex of cone-sheets (↓) or ring-dykes (↑).
**cone-sheet** (*n*) a dyke (↑) that has an outcrop in the shape of a curve and is in three dimensions shaped like a cone, dipping inwards towards a point below the Earth's surface.

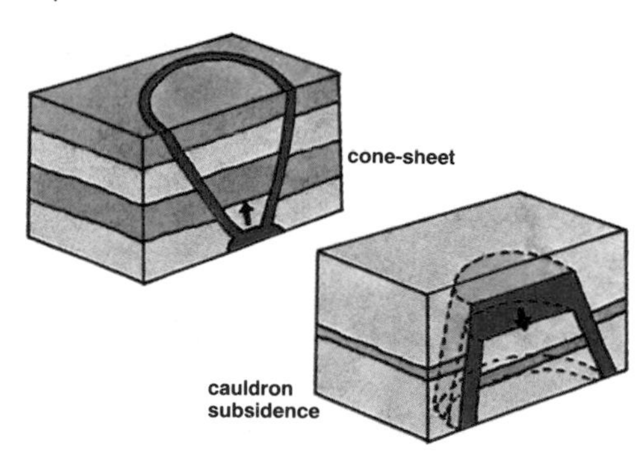


**cauldron subsidence** the descent of a large mass of rock shaped like a drum with magma (p.62) coming up from below round its sides and over it. Ring-complexes (↑) could be formed in this way.
**columnar jointing, columnar structure** a regular form of jointing (p.21) that produces six-sided pillars or columns in igneous rocks (p.62). The joints are perpendicular to the contact (p.148) between the intrusion and the country-rock (p.65). Columnar jointing is seen especially in lava flows (p.70) and sills (↑).
**minor intrusions** dykes (↑) and sills (↑).

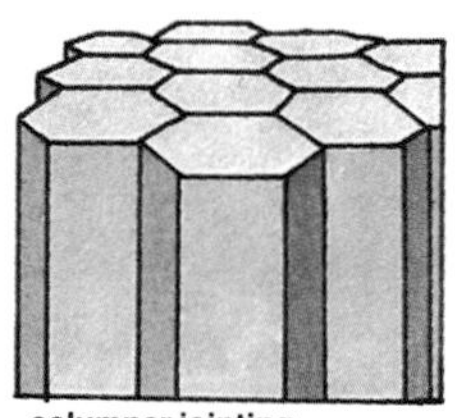
columnar jointing

**volcano** (*n*) (1) a hole (vent) or fissure (p.21) in the Earth's crust (p.9) from which molten lava (p.70), pyroclastic materials (↓), and gases come out; (2) a hill or mountain, usually shaped like a cone, built of the materials that have come out of such a vent. **volcanic** (*adj*), **volcanism, vulcanism** (*n*).

**eruption** (*n*) the sending out of volcanic material from a volcanic vent (↓) or fissure (p.21) at the Earth's surface. **erupt** (*v*), **eruptive** (*adj*).

**central eruption** a volcanic eruption (↑) from a circular vent (↓), i.e. an eruption of the ordinary kind.

**fissure eruption** a volcanic eruption (↑) from a crack or fissure (p.21) in the Earth's crust (p.9).

**volcanic vent** the hole or pipe through which a volcano sends out lava (p.70) and other igneous material (p.62) such as volcanic ash (↓).

**volcanic conduit** = volcanic vent (↑).

**volcanic cone** the heap of volcanic material – solidified lava (p.70), volcanic ash (↓), etc. – that forms round and above a volcanic vent (↑).

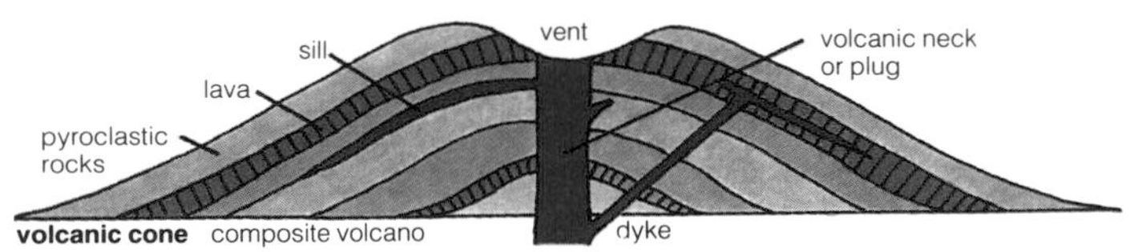


**volcanic cone** composite volcano

**composite cone** a volcanic cone (↑) formed of lavas (p.70) and pyroclastic rocks (↓).

**stratovolcano** (*n*) = composite cone (↑).

**cinder cone** a volcanic cone (↑) formed of pyroclastic rocks (↓).

**shield volcano** a volcanic cone (↑) formed by the eruption (↑) of large quantities of lava (p.70) of a type that flows easily. Shield volcanoes are large and their sides have gentle slopes.

**volcanic plug** solidified magma (p.62) in a volcanic vent (↑).

**volcanic neck** (1) a volcanic plug; (2) the pipe that joins a magma chamber to a volcanic vent (↑).

**caldera** (*n*) a large hollow, more or less circular in plan, formed by the falling in or the explosion of a volcano (↑).

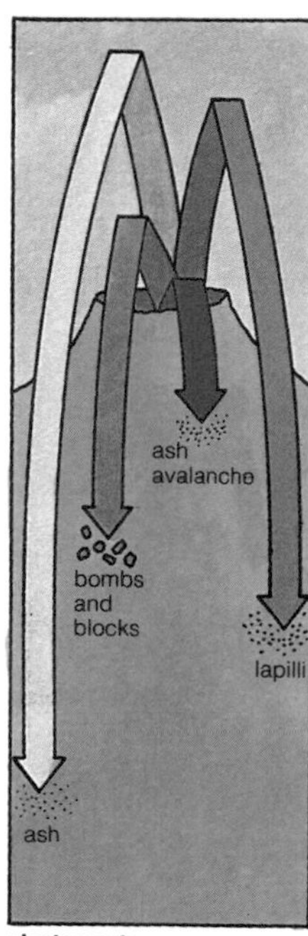

**ejectamenta**

**diatreme** (*n*) a volcanic vent (↑) that has cut through bedded rocks as a result of an explosive eruption (↑).

**extrusive** (*adj*) extrusive rocks are those rocks that have flowed out as magma (p.62) at the Earth's surface – volcanic rocks.

**pyroclastic** (*adj*) describes rocks formed of pieces of material that have been thrown into the air by volcanic action.

**ejectamenta** (*n.pl.*) solid material thrown out from a volcanic vent (↑).

**volcanic dust** very fine material in the form of particles less than 0.06 mm in diameter blown out by a volcano.

**volcanic ash** material in the form of small fragments from 0.06 to 4 mm in diameter blown out by a volcano.

**tuff** (*n*) a consolidated (p.84) volcanic ash (↑).

**lapilli** (*n.pl.*) small fragments from 4 to 32 mm in diameter blown out by a volcano.

**volcanic bomb** an irregular or long, rounded block of lava more than 32 mm in diameter thrown out by a volcano.

**volcanic block** a large mass of rock thrown out by a volcano.

**volcanic agglomerate** a pyroclastic rock (↑) made up of fragments that are 20 to 30 mm in diameter or larger.

**agglomerate** (*n*) = volcanic agglomerate (↑).

**volcanic breccia** a rock consisting of large angular pieces of volcanic rock in a matrix (p.73) of fine-grained pyroclastic (↑) material.

**pumice** (*n*) a light-coloured, glassy, vesicular (p.73) rock of acid (p.74) composition. Pumice is formed when gases pass through newly erupted rhyolitic (p.76) lava (p.70).

**scoria** (*n.pl.*) pieces of rough, highly vesicular (p.73) lava (p.70) that have been thrown out from a volcano or have been formed by the cooling of the surface of molten lava by the air. *See also p. 73.*

**spatter cone** a tower-shaped heap of lava (p.70) built up layer by layer by eruptions of lava one after the other from a small opening on the side of a volcano.

**lava** (*n*) (1) the molten material that is thrown out from a volcano; magma (p.62) that reaches the Earth's surface; (2) the rock that forms when the molten material solidifies.

**lava flow** (1) lava poured out at the Earth's surface from a vent (p.68) or fissure (p.68); (2) the solid rock formed from lava poured out in this way.

**áá** (*n*) a lava (↑) flow with a rough surface.

**pahoehoe** (*n*) a lava (↑) flow with a surface like rope and a glassy outer skin.

**lava tube** the surface of a lava flow (↑) may cool and solidify while liquid lava continues to flow beneath it in a lava tube. When the flow of lava stops the tube may be left empty.

**lava tunnel** a lava tube (↑) that is open at both ends.

lava fountain

**fire fountain** a stream of very hot lava (↑) and gas coming from a hole in the ground. The lava is basic (p.74) and flows readily. Fire fountains may reach heights of several hundred metres. A line of fire fountains may form a *fire curtain* (↓).

**fire curtain** a line of *fire fountains* (↑).

**nuée ardente** (*n. French*) (*nuées ardentes*) a white-hot cloud of gas and volcanic ash (p.69) from a volcano. A *nuée ardente* can cause great damage.

**ignimbrite** (*n*) a volcanic rock (p.68) formed from a *nuée ardente* (↑); a welded tuff (p.69).

**phreatic eruption** a volcanic eruption that takes place when water under the ground is rapidly and violently turned to steam.

**lahar** (*n*) a mud flow of ash, etc. and water on the sides of a volcano.

**pillow-lava** a lava (rock) formed by an eruption under water (typically under the sea). Pillow lavas have characteristic rounded masses that are shaped like pillows.

**fumarole** (*n*) a hole in the ground in a volcanic area from which steam and other gases (volatiles, p.18) come out. Fumaroles are characteristic of the later stages of volcanic activity.

**geyser** (*n*) a spring in a volcanic area from which hot water and steam come out from time to time.

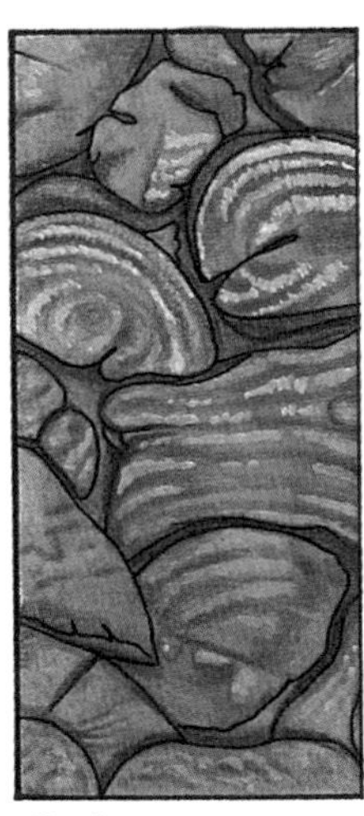

pillow lava

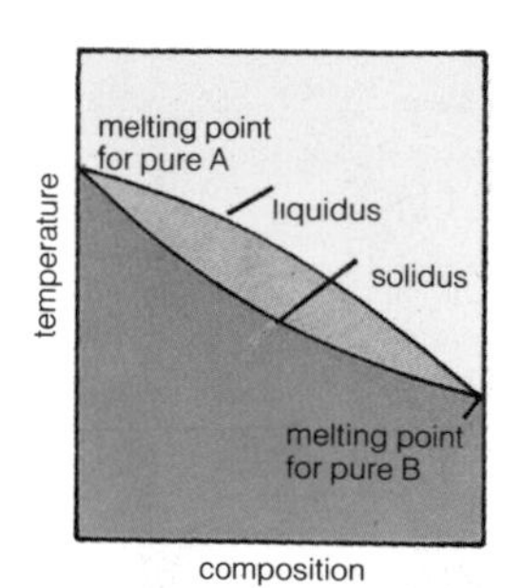


**temperature – composition**

**temperature–composition diagram** a figure drawn to show the relationship between the composition (p.15) of a mixture and temperature. Diagrams of this kind can be used to show the crystallization (p.40) of a magma (p.62) consisting of two components (p.19).

**liquidus** (*n*) a line joining points on a temperature–composition diagram (↑) of a two-component system at which the liquid phase (p.19) contains as much of a solid phase or component (p.19) as can dissolve (p.155) in it (i.e. it is *saturated*). If the temperature of the liquid falls, crystallization (p.40) will begin when the liquidus temperature is reached that corresponds to the composition of the melt.

**solidus** a line on a temperature–composition diagram that joins points above which solid and liquid are in equilibrium (p.17) and below which only the solid phase exists.

**eutectic point** a point on a temperature–composition diagram at which two components crystallize together; the lowest temperature at which a mixture of given components (p.19) will melt provided that they do not form solid solutions (p.46).

**incongruent melting** a solid may melt to form another solid phase (p.19) and a liquid, neither of which has the same chemical composition (p.15) as the original solid. This is called *incongruent melting*. For example, orthoclase (p.56) melts to form leucite (p.58) and a liquid that contains more silica (p.16) than orthoclase.

**exsolution** (*n*) unmixing. The appearance of two mineral phases (p.19) in the solid state when a solid solution (p.46) is cooled slowly to a certain temperature (called the *exsolution temperature*). One mineral then separates out. **exsolved** (*adj*).

**reaction series** (*n*) a group of minerals that are formed one after the other by chemical reaction (p.17) during the cooling of a magma (p.62).

**triangular diagram** a figure or drawing showing the chemical composition of a system (e.g. a magma) made up of three components (p.19). Each side represents a two-component system.

**ternary diagram** = triangular diagram (↑).

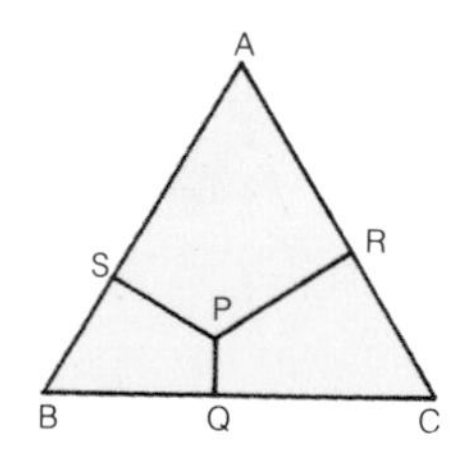


PQ represents the percentage of component A present in composition P; PR the percentage of component B; and PS the percentage of component C

**triangular diagram for system of three components**

**grain** (*n*) (1) one of the mineral particles or individual crystals (p.40) that make up a rock; (2) the texture (fine or coarse) of a rock. **grained** (*adj*).
**grain boundary** the surface that separates two grains (↑) that are next to each other in a rock.
**texture** (*n*) the relationships between the grains of minerals that make up a rock. **textural** (*adj*).
**phaneritic** (*adj*) a rock texture (↑) with crystals that are large enough to be seen by the eye without using a lens.
**aphanitic** (*adj*) a rock texture (↑) in which the crystals are too small to be seen by the eye without using a lens.
**phanerocrystalline** (*adj*) a rock texture (↑) with crystals that can be seen by the eye without using a lens.
**microcrystalline** (*adj*) a rock texture (↑) with crystals that are so small that they can be seen only by using a microscope (p.147).
**cryptocrystalline** (*adj*) a rock texture (↑) with crystals (p.40) that are so small that they cannot be seen without a powerful microscope (p.147); a finer texture than microcrystalline (↑).
**holocrystalline** (*adj*) wholly made up of crystals, without any glassy (↓) material.
**glassy** (*adj*) without any crystals: like glass in appearance. *See also* **glass** (p.77).
**hyaline** (*adj*) = glass-like in appearance (↑).
**crystalline texture** (*adj*) a rock texture (↑) in which the mineral grains are firmly locked together.
**decussate** (*adj*) a rock texture (↑) with long narrow crystals arranged in an irregular way.
**idiomorphic** (*adj*) a rock texture (↑) in which the mineral grains (↑) show more or less complete crystal forms.
**panidiomorphic** (*adj*) a rock texture (↑) in which most of the mineral grains (↑) are idiomorphic (↑).
**hypidiomorphic** (*adj*) a rock texture (↑) in which the mineral grains (↑) show some crystal form.
**allotriomorphic** (*adj*) a rock texture in which the mineral grains (↑) do not show crystal form (p.40).

**glassy texture** (obsidian)

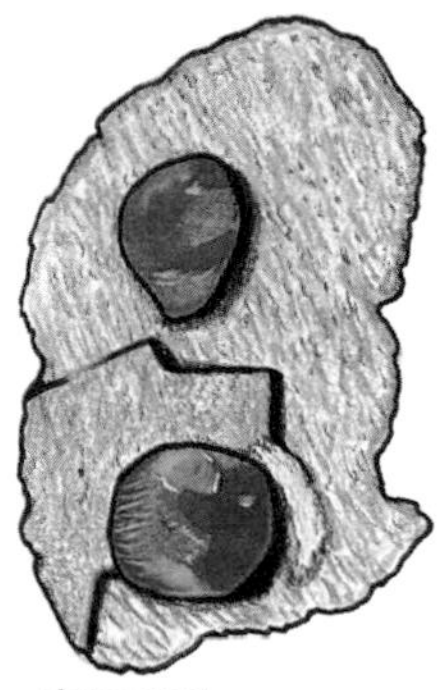

**phenocryst**
garnet in feldspar

**porphyritic** (*adj*) a rock texture (↑) with large crystals or phenocrysts (↓) in a fine-grained groundmass (↓).

**phenocryst** (*n*) a larger crystal in a groundmass (↓) of finer grain (↑). *See also* **porphyritic** (↑).

**groundmass** (*n*) the material of finer grain (↑) that forms the main part of an igneous rock with larger crystals.

**matrix** (*n*) material of finer grain (↑) between larger grains.

**equigranular** (*adj*) a rock texture (↑) in which all the mineral grains (↑) are of about the same size.

**xenocryst** (*n*) a crystal (p.40) that was not formed in the rock in which it is found.

**spherulite** (*n*) a small round mass of crystals formed when a volcanic (p.68) glass is devitrified (p.63). **spherulitic** (*adj*).

**amygdaloidal** (*adj*) a texture (↑) shown by lavas (p.70) in which holes, more or less round in shape, were formed when gas was given off by the magma. These holes are called *amygdales* or *amygdules*. They may later be filled with a mineral; some geologists use the word only for rocks in which the amygdales have been filled in this way. *See also* **scoriaceous** (↓).

**vesicle** (*n*) a more or less round hole in a lava (p.70) which has been formed by gas from the magma (p.62). **vesicular** (*adj*). *See also* **amygdaloidal** (↑).

**scoriaceous** (*adj*) a word used for an amygdaloidal (↑) texture (↑) in which the holes (amygdales) are empty.

**ophitic** (*adj*) a rock texture in which larger crystals of pyroxene (p.57) contain euhedral or subhedral (p.45) crystals of plagioclase (p.56). *See also* **poikilitic** (↓), **symplectic** (↓).

**poikilitic** (*adj*) a texture in which small crystals, facing in various directions, are inside a larger crystal. *See also* **ophitic** (↑), **symplectic** (↓).

**intergrowth** (*n*) a relationship in which crystals of different minerals are firmly locked together.

**symplectic** (*adj*) a rock texture in which there is an intergrowth (↑) of two different minerals, e.g. ophitic texture (↑), poikilitic texture (↑).

**saturation** (*n*) the degree to which a rock is saturated. In petrology (p.62) an *oversaturated* rock is one that contains free silica, $SiO_2$; that is, silica that is not chemically combined. The free silica may appear as quartz (p.55). An *undersaturated* rock contains only minerals that do not appear when free silica is present (such as olivine (p.58)). **saturated** (*adj*).

**oversaturated** (*adj*) *see* **saturation** (↑).

**undersaturated** (*adj*) *see* **saturation** (↑).

**acid** (*adj*) in petrology an acid igneous rock is one that has 10 per cent or more free quartz; granite and rhyolite (p.76) are examples. The use of the word 'acid' in this way has its origin in an earlier view of silicates as compounds of silica ($SiO_2$) with oxides of metals. Silica was then thought of as playing the part of an acid. *See also* **basic** (↓).

**basic** (*adj*) in petrology a basic igneous rock is one that contains no quartz and has feldspars (p.56) with more calcium than sodium; basalt (p.77) and gabbro (p.76) are examples. Basic rocks contain 45 to 55 per cent silica. *See also* **acid** (↑).

**intermediate** (*adj*) an intermediate igneous rock is one with less than 10 per cent quartz and with plagioclase feldspar containing 50–70% albite (p.55) or alkali feldspar (p.56), or both. *See also* **acid** (↑) *and* **basic** (↑).

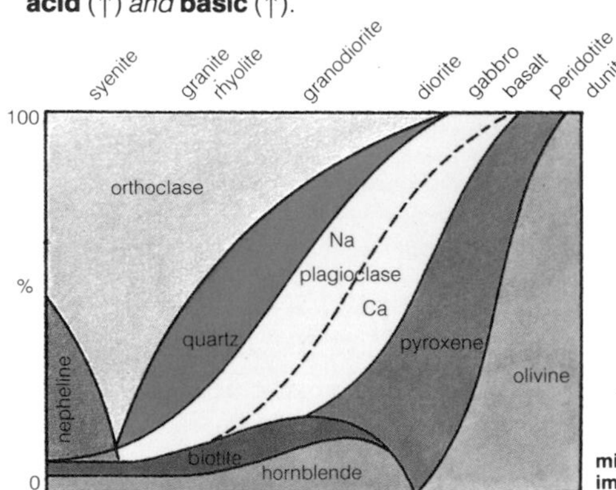


**mineral composition of important rock types**

**ultrabasic, ultramafic** (*adj*) ultrabasic or ultramafic igneous rocks are those composed essentially of ferromagnesian minerals (p.55); quartz, feldspars, and feldspathoids (pp.55, 56, 58) are absent.

**ultramafic**, *see* **ultrabasic** (↑).

**alkaline** (*adj*) alkaline igneous rocks are those in which the feldspar (p.56) is mainly sodium-bearing or potassium-bearing. ('Alkaline' here does not mean the opposite of 'acid'.) The word is also used with the same meaning for minerals. *See also* **calc-alkaline** (↓) *and* **alkali** (p.16).

**calc-alkaline** (*adj*) calc-alkaline igneous rocks are those in which the feldspar (p.56) is rich in calcium. The word is also used for minerals that are rich in calcium. *See also* **alkaline** (↑).

**essential mineral** a mineral (p.44) that must be present in an igneous rock (p.62), if only in small amounts, for it to be given a particular name, e.g. quartz and alkali feldspar are essential minerals in granite. *See also* **accessory mineral** (↓).

**accessory mineral** a mineral (p.44) whose presence or absence does not affect the name given to an igneous rock (p.62). *See also* **essential mineral** (↑).

**monomineralic** (*adj*) consisting of only one mineral.

**mafic** (*adj*) a word used for the ferromagnesian minerals (p.55) in igneous rocks. *See also* **felsic** (↓).

**felsic** (*adj*) a word used for the light-coloured minerals in igneous rocks: feldspars, feldspathoids, quartz, etc. *See also* **mafic** (↑).

**leucocratic** (*adj*) describes igneous rocks consisting mainly of felsic (↑) minerals. *See also* **melanocratic** (↓).

**basalt, andesite, granite:** basalt is melanocratic, granite is leucocratic

**melanocratic** (*adj*) describes igneous rocks consisting mainly of dark minerals. *See also* **leucocratic** (↑).

**petrographic province** an area in which a group of igneous rocks formed at about the same time are similar in petrographic (p.62) character and have a common origin, e.g. the British–Icelandic Tertiary igneous province.

**granite** (*n*) a coarse-grained acid plutonic (p.64) rock consisting essentially of quartz (p.55), alkali feldspar (p.56), and mica (p.55). The silica content is about 70 per cent. Granite is the most common intrusive rock. **granitic** (*adj*).

**granodiorite** (*n*) a coarse-grained acid plutonic rock (p.64) containing quartz (p.55), plagioclase (p.56), orthoclase (p.56), and ferromagnesian minerals (p.55), chiefly biotite (p.55) and hornblende (p.57). **granodioritic** (*adj*).

granite

**syenite** (*n*) a coarse-grained intermediate (p.74) igneous rock consisting essentially of alkali feldspar (p.56) with hornblende (p.57), biotite, or other mafic minerals. Most syenites are plutonic (p.64). **syenitic** (*adj*).

**diorite** (*n*) a coarse-grained plutonic rock of intermediate composition (p.74) consisting essentially of intermediate plagioclase (oligoclase-andesine) (p.56) hornblende, and biotite or pyroxene as ferromagnesian minerals (p.55). A diorite may contain up to 10 per cent quartz (p.55) and some alkali feldspar (p.56). **dioritic** (*adj*).

**gabbro** (*n*) a coarse-grained plutonic rock of basic composition (p.74) consisting of calcic plagioclase (labradorite–andesine) (p.56) and pyroxene (p.57). Many gabbros also contain olivine. **gabbroic** (*adj*).

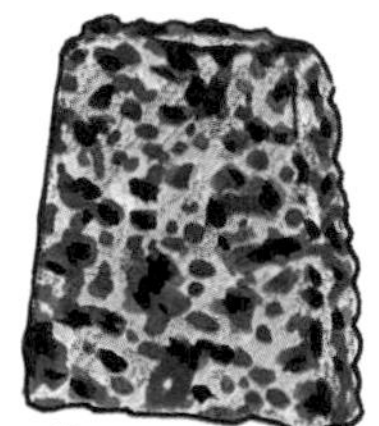
gabbro

**rhyolite** (*n*) an acid (p.74) volcanic rock, fine-grained or glassy (p.72), approximately equivalent to granite (↑). **rhyolitic** (*adj*).

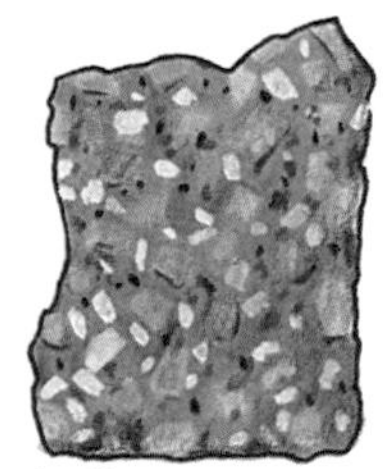
rhyolite

**dacite** (*n*) a fine-grained rock equivalent to a granodiorite (↑).

**andesite** (*n*) a fine-grained intermediate rock (p.74) composed essentially of plagioclase (oligoclase–andesine) (p.56) with biotite (p.55), hornblende (p.57), or pyroxene (p.57). Andesites are more or less equivalent in chemical and mineralogical composition to diorites (↑). Andesites occur as extrusive rocks (p.69) and as dykes and sills (pp.66, 67). **andesitic** (*adj*).

**microgranite** (*n*) a medium-grained acid igneous rock (p.62) similar in chemical and mineral composition to a granite (↑).

**basalt** (*n*) a fine-grained basic igneous rock (p.74) consisting essentially of calcic plagioclase (p.56) and pyroxene (p.57), usually augite, with or without olivine (p.58). Basalt is more or less equivalent to gabbro (↑). Basalts occur mainly as lavas (p.70). They may be porphyritic, vesicular, or amygdaloidal (p.73). **basaltic** (*adj*).

**plateau basalt** a basalt (↑) that has been poured out in large quantities from a fissure eruption (p.68). Lava-flows of this kind form *plateaux* – wide areas of land at a high level.

**tholeiite** (*n*) a basalt (↑) in which the plagioclase (p.56) is labradorite or bytownite and the pyroxene (p.57) is an Mg/Fe variety. **tholeiitic** (*adj*).

**pitchstone** (*n*) a glassy acid igneous rock (p.74) with a dull lustre (p.47) like pitch (a black sticky material that is used to fill the cracks between the boards of boats). Pitchstone occurs as dykes and sills (p.66).

obsidian

**obsidian** (*n*) a glassy volcanic rock with a composition equivalent to granite or rhyolite (↑). It is black in colour and has a conchoidal fracture (p.44).

**felsite** (*n*) a fine-grained acid igneous rock (p.74), light in colour. Quartz and feldspar (p.56) are the chief minerals. Felsites occur as dykes and veins (pp.67, 145). **felsitic** (*adj*).

**trachyte** (*n*) a fine-grained intermediate rock (p.74) composed of alkali feldspar (p.56) and ferromagnesian minerals (p.55). **trachytic** (*adj*).

**phonolite** (*n*) an undersaturated (p.74) trachyte (↑) containing the feldspathoid nepheline (p.58), alkali feldspar (usually sanidine) (p.56), and a ferromagnesian mineral (p.55). Phonolites are the fine-grained equivalents of nepheline-syenites (↑). **phonolitic** (*adj*).

**MORB** short for mid-ocean ridge basalt; a basalt (↑) erupted at a mid-oceanic ridge (p.35).

**glass** (*n*) an amorphous (p.44) rock that contains no crystalline (p.44) material, e.g. obsidian (↑). Glasses are formed when a magma (p.62) is cooled very rapidly.

**dolerite** (*n*) a medium-grained hypabyssal rock (p.64) with calcic plagioclase (p.56) and pyroxene (p.57) as essential minerals. Mineralogically equivalent to gabbro (p.76) and basalt (p.77). Dolerites are very common as minor intrusions. Ophitic texture (p.73) is common. **doleritic** (*adj*).

**diabase** (*n*) a word used by American geologists for dolerite (↑).

**monzonite** (*n*) a coarse-grained igneous rock between syenite (p.76) and gabbro (p.76), with more or less equal amounts of potassium feldspar (p.56) and plagioclase (oligoclase–andesine) (p.56). Pyroxene, biotite, or hornblende (p.57) may also be present. Monzonites range from acid to basic (p.74) in composition. **monzonitic** (*adj*).

**syenodiorite** (*n*) = monzonite (↑).

**lamprophyre** (*n*) a dark-coloured basic igneous rock with phenocrysts (p.73) of dark mica, augite, hornblende, or olivine (p.58) in a groundmass (p.73) of feldspar, usually alkali feldspar (p.56). Lamprophyres occur typically as dykes (p.67). **lamprophyric** (*adj*).

**porphyry** (*n*) a word used, generally with the name of a mineral, to describe rocks with phenocrysts (p.73) in a fine-grained groundmass, i.e. rocks with a porphyritic texture (p.73). Porphyries are usually hypabyssal rocks (p.64).

**ophiolite** (*n*) ophiolites are a group of basic and ultrabasic rocks that occur as lavas and minor intrusions (pp.64, 70) in geosynclinal areas (p.132). The word 'ophiolite' is also used for the same rocks when they are metamorphosed (p.90). **ophiolitic** (*adj*).

**ophiolite assemblage** the occurrence together of deep-sea sediments (p.80), basaltic pillow lavas (p.70), basaltic dykes (p.67), gabbros (p.76), and peridotites (↓). These assemblages may be fragments of oceanic crust (p.9).

**spilite** (*n*) an altered basalt (p.77) in which albite replaces plagioclase feldspar (p.56), and chlorite (p.61) replaces augite and olivine (p.58). Spilites occur as pillow-lavas (p.70) with geosynclinal (p.132) sediments. **spilitic** (*adj*).

**dolerite**

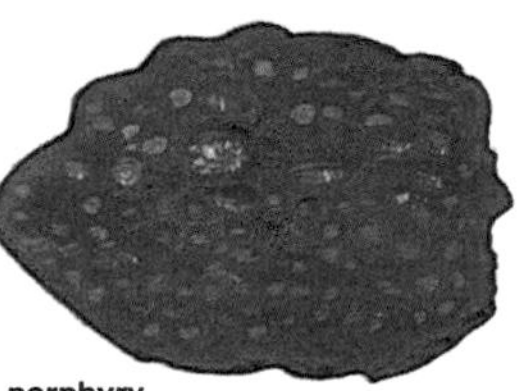
**porphyry**

**peridotite** (*n*) a dark-coloured coarse-grained ultrabasic (p.75) igneous rock consisting largely of olivine (p.58), with or without pyroxene or hornblende (p.57). Quartz and feldspar (p.56) are absent. **peridotitic** (*adj*).

**dunite** (*n*) an ultrabasic igneous rock composed almost entirely of olivine (p.58).

**pyroxenite** (*n*) a dark-coloured coarse-grained ultrabasic (p.75) igneous rock consisting essentially of one or more pyroxenes (p.57).

**anorthosite** (*n*) a coarse-grained plutonic rock consisting largely of plagioclase feldspar (andesine to labradorite) (p.56); the other minerals present are those found in gabbros (p.76). **anorthositic** (*adj*).

**hornblendite** (*n*) a coarse-grained igneous rock consisting largely of hornblende (p.57).

**perknite** (*n*) the perknites are ultrabasic rocks (p.75) consisting largely of ferromagnesian minerals (p.55) with the exception of olivine (p.58). They include pyroxenites (↑) and hornblendites (↑).

**pegmatite** (*n*) an igneous rock of coarse or very coarse grain. The crystals in a pegmatite may be up to a metre or even more in length. Granite-pegmatites, which are usually simply called pegmatites, consist of quartz and feldspar (p.56). Coarse-grained varieties of other plutonic rocks are called by the name of the equivalent rock, e.g. gabbro-pegmatite. Pegmatites occur as dykes (p.67) and veins (p.145). They are formed towards the end of the crystallization of a magma. **pegmatitic** (*adj*).

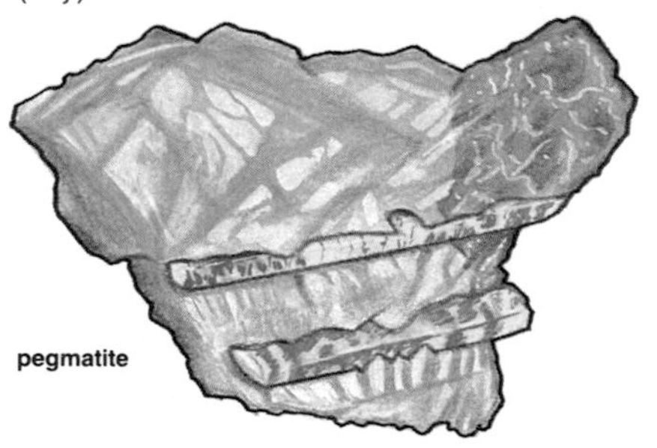

**pegmatite**

**sediment** (*n*) material that has been deposited (↓) in water (e.g. on the sea floor or on the bed of a lake), having settled after being in suspension (p.24); any material that has been obtained from earlier rocks by denudation (p.32). In the wider sense, sediment includes material deposited by ice and the wind or chemically precipitated (p.18) in water, together with material from plants and animals. **sedimentary** (*adj*).

**sedimentation** (*n*) the process of forming sediments (↑).

**sedimentology** (*n*) the study of sediments (↑) and their formation.

**deposit** (*n*) anything that is laid down; a sediment (↑); minerals precipitated (p.18) from solution (p.159) in veins (p.145) and ore bodies (p.145). **deposited** (*adj*), **deposit** (*v*).

**deposition** (*n*) the laying down of material that may later form a rock; sedimentation (↑); the precipitation (p.18) of minerals from solution (p.159).

**bed** (*n*) a layer of sedimentary (↑) rock that is marked off above and below by surfaces that can be seen and is made up of material that is the same in all parts. **bedding** (*n*).

**stratum** (*n*) (*strata*) = bed (↑).

**bedding-plane** (*n*) a surface in a sedimentary (↑) rock that is parallel to the original surface on which the sediment was deposited.

**lamina** (*n*) (*laminae*) a thin layer (less than 10 mm thick) in a sedimentary rock that can be separated from the material above and below it. **laminated** (*adj*).

**lamination** (*n*) the presence or formation of laminae (↑).

**stratification** (*n*) the presence of layers or beds (↑) in a sedimentary rock (↑).

**stratified** (*adj*) formed of layers or beds (↑).

**unstratified** (*adj*) not stratified (↑).

**fissile** (*adj*) easily divided up along closely spaced parallel surfaces.

**sorting** (*n*) the degree to which the particles that make up a sediment (↑) or other material are alike in some respect, e.g. their size or shape. **sorted** (*adj*).

**bedding plane**

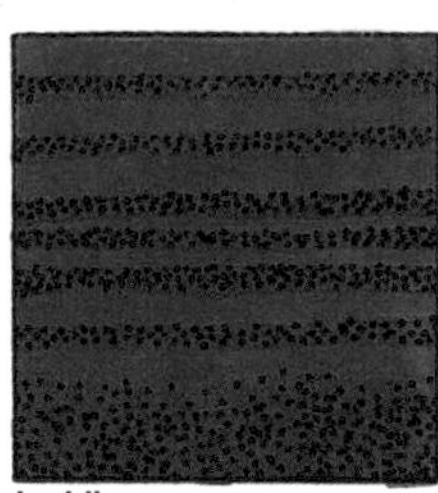
**bedding**

**environments of the seas and oceans**

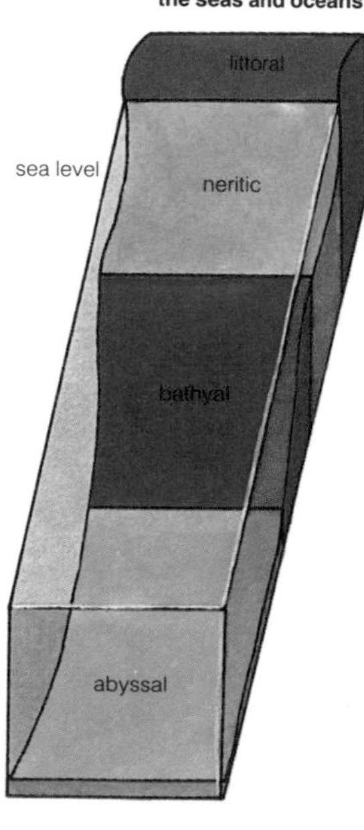


**sedimentary environment** the conditions under which a sediment (↑) is deposited (↑); e.g. the depth and temperature of the water, the strength and direction of the currents. These can vary widely, and they affect the texture, composition, and structure of the sediments that are formed.

**depositional environment** = sedimentary environment (↑).

**marine environments** these include the *littoral* (p.37) *environment*, the *neritic* (p.34) *environment*, the *bathyal* (p.34) *environment*, and the *abyssal* (p.34) *environment*.

**paralic** (*adj*) relating to sediments deposited on the coast in shallow water. Marine (p.34) and non-marine sediments may both be formed here. *See also* **limnic** (↓).

**terrestrial** (*adj*) of the land, formed on the land; not marine (p.34).

**limnic** (*adj*) relating to sediments deposited in fresh-water lakes. *See also* **lacustrine** (↓).

**lacustrine** (*adj*) of, or formed in or by, a lake. *See also* **limnic** (↑).

**paludal** (*adj*) of low wet land (swamp, marsh).

**lagoonal** (*adj*) relating to areas of shallow water on the coast, usually between a barrier island (p.38) and the mainland.

**fluvial** (*adj*) relating to rivers.

**estuarine** (*adj*) relating to the part of a river nearest to the sea, where tides, i.e. the regular rise and fall of the sea, occur and where fresh and salt water mix. The sediments deposited in estuaries are fine silts, clays, and muds.

**deltaic** (*adj*) relating to a delta (p.26). Sediments laid down in a delta typically show cross bedding (p.82).

**anaerobic** (*adj*) without oxygen. Sediments formed under anaerobic conditions are typically black muds rich in organic (p.17) material and sulphides (p.16).

**allothigenous, allothigenic** (*adj*) refers to the part of a sediment (↑) that has been transported (p.21) from some other place to the area in which it was deposited (↑).

**allogenic** (*adj*) = allothigenous (↑).

**graded bedding** bedding (p.80) in which the largest particles are at the bottom of a unit and the smallest particles are at the top.
**cross bedding** original bedding (p.80) in which the bedding-planes (p.80) are at an angle to the main surface on which the sediments were deposited (p.80). **cross bedded** (*adj*).
**current bedding** = cross bedding (↑).

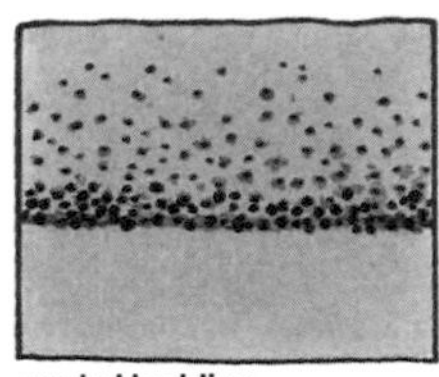
graded bedding

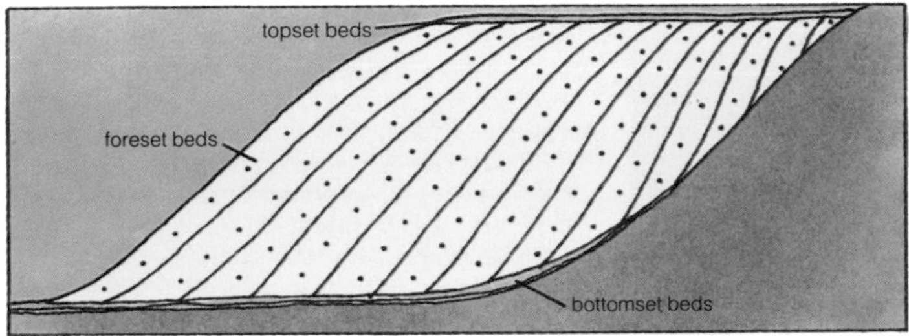

cross-(current-) bedding

**dune bedding** cross bedding (↑) of relatively large size. **dune-bedded** (*adj*).
**topset beds** horizontal strata (p.80) at the top of sediments deposited (p.80) in a delta (p.26).
**bottomset beds** horizontal strata (p.80) at the bottom of sediments deposited (p.80) in a delta (p.26).
**foreset beds** steeply sloping beds (p.80) in sediments deposited (p.80) in a delta (p.26).
**washout** (*n*) a gap in a bed (p.80) that has been filled with later sediments (p.80). Formed when a stream flows across sediments soon after they have been deposited. The sediments in the course of the stream will then be removed and the space will later be filled by other material.
**scour-and-fill** the process of cutting a channel in a sediment and filling it in again. *See also* **washout** (↑).
**convolute bedding** a structure in which the laminae (p.80) of a sediment are bent into folds, which are cut off by the beds above.
**palaeocurrent** (*n*) a flow of water that took place at some time in the geological past while a sediment was being deposited (p.80).

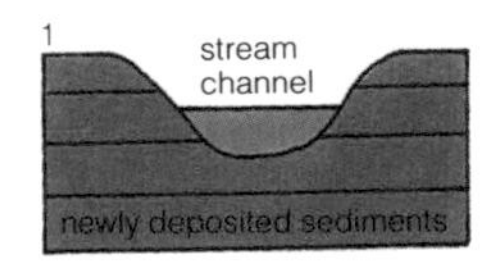


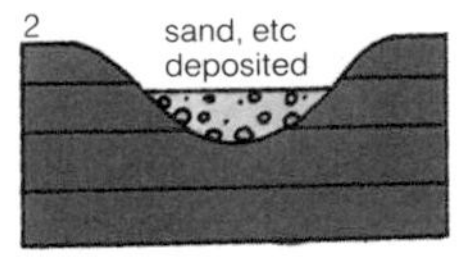


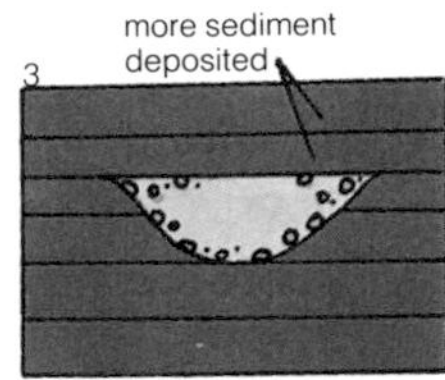

formation of washout

**angular** (*adj*) with sharp edges and corners; showing little sign of wear.

**subangular** (*adj*) with edges and corners rounded off to some degree; showing signs of wear. *See also* **angular** (↑).

**subrounded** (*adj*) with edges and corners rounded to smooth curves but with the original shape of the grain still to be seen.

**rounded** (*adj*) with edges and corners rounded to smooth curves.

**well-rounded** (*adj*) worn to a completely smooth curved shape without any sign of the original shape of the particle.

**sedimentary structures** structures formed while a sediment (p.80) was being deposited (p.80) or very soon after it was deposited.

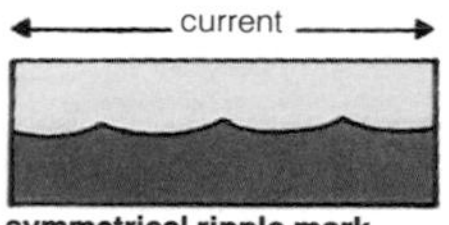


**symmetrical ripple mark**

**ripple mark** wave-like marks formed by the movement of water or of the air over the surface of a newly deposited sediment. The two main types of ripple mark are *oscillation ripples*, which are symmetrical, and *current ripples*, which are not symmetrical.

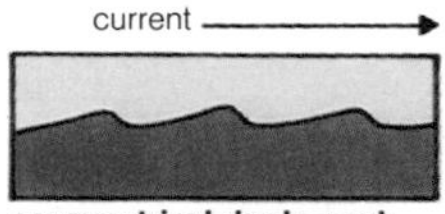


**asymmetrical ripple mark**

**swash marks** patterns made in the sand of a beach by the movement of water after a wave has broken. These patterns are usually curved.

**crescent marks** marks formed by the washing away by water of the sediment round a pebble, a shell, or other object on the beach as the sea flows back after each wave.

**mud cracks**

**rill marks** branching patterns of small channels formed by water running down a beach.

**mud cracks** the cracks produced when wet mud dries in the air. A pattern of many-sided cracks results, typical of dried up shallow lakes.

**flow cast** a swelling formed when sediment is deposited on a soft material that can flow under pressure.

**load cast** = flow cast (↑).

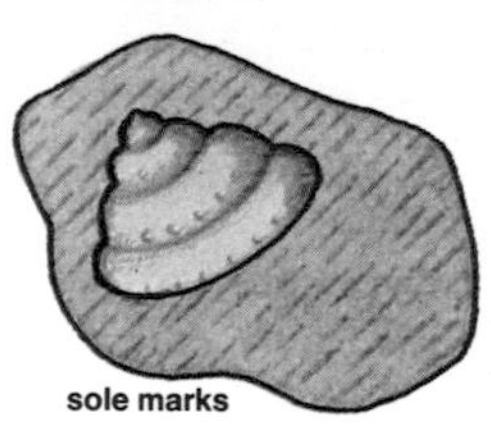

**sole marks**

**sole marks, sole markings** marks on what was originally the under surface of a bed (p.80). The bed must be different lithologically (p.85) from the bed below it for sole marks to be seen. Types of sole mark include various marks produced by animals and objects moving on the surface on which the sediment was deposited (p.80).

**compaction** (*n*) the process in which a sediment is reduced in volume by pressure, e.g. by the weight of material above it. **compact** (*v*), **compacted** (*adj*).

**consolidation** (*n*) the process of forming a solid mass from loose or liquid material, e.g. the formation of a firm rock from loose sediment. **consolidate** (*v*), **consolidated** (*adj*).

**unconsolidated** (*adj*) not consolidated (↑); loose.

**lithification** (*n*) the processes by which unconsolidated sediments become consolidated (↑) into rocks. **lifthify** (*v*), **lithified** (*adj*).

**lithifaction** = lithification (↑).

**induration** (*n*) the process of making hard, e.g. by heat or by pressure. **indurate** (*v*), **indurated** (*adj*).

**cement** (*n*) material between the particles of a sedimentary rock that holds them together.

**cementation** (*n*) the deposition of cement (↑) between the particles of a sediment to form a solid rock.

**impregnation** (*n*) the filling-in of the pore spaces (↓) in a rock by minerals or the replacement (p.159) of pore material. Impregnation generally relates to an event that takes place after cementation (↑). **impregnate** (*v*), **impregnated** (*adj*).

**pore space** the space between the particles that make up a rock.

**porosity** (*n*) the proportion of empty space in a rock.

**porous** (*adj*) containing pore spaces (↑).

**diagenesis** (*n*) the changes in mineral composition and texture that take place in a sediment after it has been deposited, except those that occur at depth. **diagenetic** (*adj*).

**authigenic** (*adj*) minerals (p.44) formed during diagenesis (↑) are called *authigenic*.

**concretion** (*n*) a mass of round or irregular shape in a sediment, formed either during deposition (p.80) or during diagenesis (↑).

**nodule** (*n*) a concretion (↑) of rounded shape.

**Neptunean dyke** a sheet of sediment (p.80) filling a crack in another rock.

**sandstone dyke** = Neptunean dyke (↑).

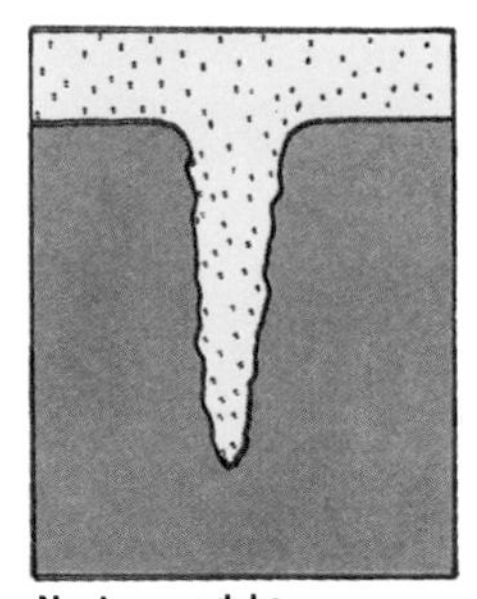

**Neptunean dyke**

**lithology** (*n*) the general character of a rock, or a sedimentary formation, more particularly as seen in exposures (p.122) and hand specimens (p.147). **lithological** (*adj*).

**rudite** (*n*) a sedimentary rock with an average grain size greater than 2 mm, e.g. conglomerates and breccias (p.87). **rudaceous** (*adj*).

**psephite** (*n*) = rudite (↑). **psephitic** (*adj*).

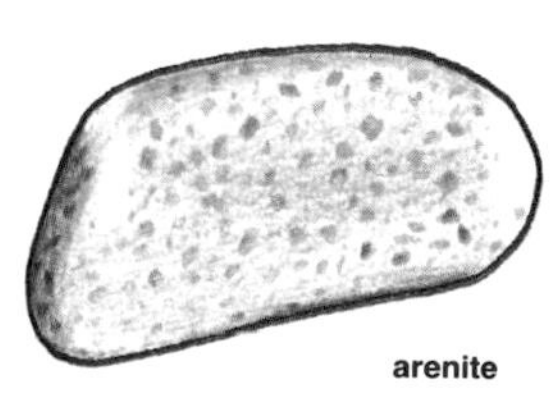

arenite

**arenite** (*n*) a sedimentary rock with a grain size from 1/16 to 2 mm; a sandstone. **arenaceous** (*adj*). *See also* **sandstone**, (p.87).

**lutite** (*n*) a sedimentary rock of any composition with particles between 1/256 and 1/16 mm in diameter; a rock composed of mud. **lutaceous** (*adj*).

**argillite** (*n*) a sedimentary rock with an average grain size less than 1/16 mm; clays, silts, mudstones, etc.; a hard mudstone. **argillaceous** (*adj*). *See also* **clay** (p.88).

**clast** (*n*) a piece of a sedimentary (p.80) rock formed by the breaking-up of a larger mass.

**clastic** (*adj*) describes sediments made up of fragments produced by the breaking-up of earlier rocks.

**detrital** (*adj*) = clastic (↑).

**organic** (*adj*) as applied to sediments, 'organic' refers to the remains of plants and animals; e.g. coral reefs and coal are organic sediments.

**evaporite** (*n*) a sediment formed by the *evaporation* (i.e. the drying up) of a body of water containing a chemical compound in solution, e.g. salt deposits, gypsum.

**terrigenous** (*adj*) applied to sediments deposited on the sea bed that contain material that has come from the land.

**orthochemical** (*adj*) refers to materials in sedimentary rocks that have been chemically precipitated (p.18) within the area in which the rock was deposited and show no sign of having been transported after their deposition.

**allochemical** (*adj*) refers to constituents of sedimentary rocks that have been chemically precipitated (p.18) within the area in which the rock was deposited but have since been transported elsewhere; e.g. shell fragments.

**limestone** (*n*) a sedimentary rock (p.80) containing more than 50 per cent of calcium carbonate, $CaCO_3$. Limestones can be of fresh-water or marine origin. The material of which they are composed may be chemically precipitated (p.18), organic (p.85), or detrital (p.85).

**calcareous** (*adj*) containing calcium carbonate, $CaCO_3$.

**chalk** (*n*) a very pure limestone, white in colour. The word is also used, with a capital C, as a stratigraphical name for a division of the Tertiary (p.115).

**oolith** (*n*) ooliths are small round particles that make up a sedimentary rock. In the mass they look like the roe (eggs) of a fish. Ooliths have a regular structure. They are usually calcareous (↑). **oolitic** (*adj*).

**oolite** (*n*) an oolitic (↑) limestone (↑); a limestone composed of ooliths.

**pisolith** (*n*) a large oolith (↑), 3–6 mm in diameter.

**pisolite** (*n*) a coarse-grained oolite (↑) containing ooliths 3–6 mm in diameter.

**micrite** (*n*) very fine-grained calcite (p.51) forming the matrix (p.73) of a limestone (↑). *See also* **sparite** (↓).

**sparite** (*n*) coarse-grained calcite (p.51) forming the matrix (p.73) of a limestone (↑).

**dolomitization** (*n*) the conversion of calcium carbonate, $CaCO_3$, in a rock to dolomite, $CaMg(CO_3)_2$ (p.51). The process may be partial or complete. Dolomitization is usually caused by metasomatism (p.90). **dolomitized** (*adj*).

**dedolomitization** (*n*) the conversion of dolomite, $CaMg(CO_3)_2$ (p.51), to calcite, $CaCO_3$, in a rock. This may be caused by thermal metamorphism (p.90) or diagenesis (p.84).

**chert** (*n*) a dense, hard, siliceous rock (↓); cryptocrystalline (p.44) silica, $SiO_2$; a form of quartz (p.55). Chert occurs as beds (bedded chert) and as nodules (p.84) in limestones (↑) and shales (p.88).

**flint** (*n*) a variety of chert (↑) occurring as irregular nodules (p.84), especially in chalk (↑).

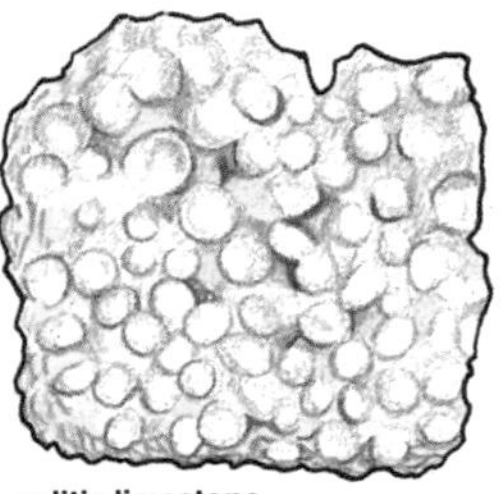

**oolitic limestone**

**sandstone** (*n*) a clastic (p.85) arenaceous rock (p.85) consisting of fragments from 1/16 to 2 mm in diameter. The fragments are rounded to subrounded (p.83) in shape. Most sandstones are composed mainly of quartz grains (p.55).

**grit** (*n*) an arenaceous rock (p.85) in which the particles are angular to subangular (p.83). *See also* **sandstone** (↑).

**quartzite** (*n*) a quartz sandstone (↑) with a quartz (p.55) cement (p.84).

**orthoquartzite** = quartzite (↑).

**siliceous** (*adj*) containing silica, $SiO_2$; containing large amounts of quartz (p.55).

**arkose** (*n*) an arenaceous rock (p.85) composed mainly of fragments of quartz (p.55) and feldspar (p.56). Commonly red or pink in colour.

**greywacke, graywacke** (*n*) a dark-coloured arenaceous rock (p.85) consisting of angular and subangular (p.83) fragments of various sizes from fine to coarse; an impure sandstone (↑). Greywackes are formed in mobile belts (p.132); they show sedimentary structures (p.83).

**greensand** (*n*) a sandstone (↑) containing glauconite (p.55).

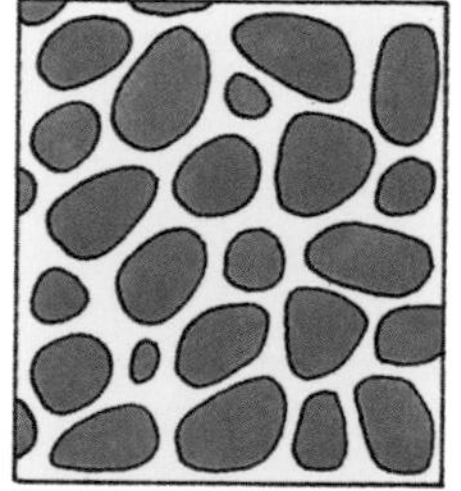
conglomerate

**conglomerate** (*n*) a rudaceous rock (p.85) composed of rounded or subrounded (p.83) fragments. The fragments may be of any rock and may be from a few millimetres to several centimetres in diameter.

**breccia** (*n*) a clastic (p.85) sedimentary rock composed of angular (p.83) fragments mixed with finer material. *See also* **conglomerate** (↑).

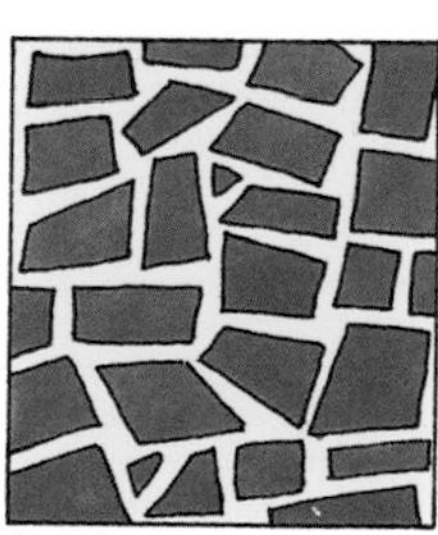
breccia

**sand** (*n*) detrital (p.85) material consisting of particles from 1/16 to 2 mm in diameter. *See also* **sandstone** (↑).

**gravel** (*n*) (1) a sediment with grains from 2 to 4 mm across; (2) loose detrital (p.85) material of a range of sizes.

**pebble** (*n*) a smooth rounded (p.83) piece of rock between 4 mm and 64 mm in diameter.

**cobble** (*n*) a rounded (p.83) piece of rock between 64 mm and 256 mm in diameter.

**boulder** (*n*) a large rounded piece of rock more than 256 mm in diameter.

**clay** (*n*) an argillaceous rock (p.85) with particles less than 1/256 mm in diameter. Clays are plastic when wet and show no bedding (p.80).

**shale** (*n*) an argillaceous rock (p.85) with particles less than 1/256 mm in diameter showing well-marked bedding. Shale is fissile (p.80).

**mudstone** (*n*) an argillaceous rock (p.85) like a shale (↑) but not fissile (p.80).

**marl** (*n*) a calcareous (p.86) mudstone (↑).

**silt** (*n*) a sediment with particles from 1/16 to 1/256 mm in diameter.

**siltstone** (*n*) a consolidated silt (↑); an argillaceous rock (p.85) like a mudstone (↑) but with particles of silt (↑) grade.

**ferruginous** (*adj*) containing iron. Ferruginous sediments may contain large amounts of iron compounds, e.g. bedded siderites and sedimentary haematite (p.48).

**red beds** a general term for sedimentary rocks that contain red iron compounds. Red beds are usually formed under very dry continental conditions.

**ironstone** (*n*) a rock, usually sedimentary, that is made up largely of chemical compounds of iron.

**clay-ironstone** (*n*) an argillaceous (p.85) ironstone. The iron is usually present as siderite (p.51), iron carbonate ($FeCO_3$). Clay-ironstone occurs in beds (p.80) and as concretions (p.84).

**black-band ironstone** a clay-ironstone (↑) containing coal-like material. It can be burnt to produce iron.

lignite

cannel coal

**carbonaceous** (*adj*) containing carbon. Carbonaceous rocks include coal (↓) and lignite (↓).

**peat** (*n*) an accumulation of vegetable matter at the Earth's surface that has partly decomposed.

**lignite** (*n*) a carbonaceous (↑) rock of a kind between peat (↑) and bituminous coal (↓); called 'brown coal'.

**coal** (*n*) plant remains that have been changed physically and chemically to form a hard, black substance that can be burned.

**bituminous coal** ordinary coal (↑) such as is used in the home for burning. It contains a relatively small proportion of carbon.

**anthracite** (*n*) a coal (↑) containing more than 90% of carbon.

**bitumen** (*n*) a mineral composed of compounds of carbon and hydrogen (hydrocarbons) that looks like tar. Bitumen is found as a thick liquid or as a solid that breaks easily.

**asphalt** (*n*) a brown or black almost solid hydrocarbon rather like bitumen (↑). Asphalt occurs naturally (e.g. in Trinidad) and is also formed in the production of petroleum (p.144).

**cannel coal** a fine-grained coal, dull in colour.

**jet** (*n*) a type of cannel coal (↑) or lignite (↑), black in colour and hard. It can be polished and made into ornaments.

**sub-bituminous coal** a coal (↑) of rank (↓) between bituminous coal (↑) and lignite (↑). Most coals of this type are of Mesozoic and Tertiary age (p.115).

**coal seam** a bed of coal.

**coal measures** strata (p.80) with coal seams (↑), or the coal seams themselves.

**rank (of coal)** the rank of a coal is a measure of the amount of carbon that it contains. Thus anthracite (↑), with a high proportion of carbon, is of high rank; bituminous coal, with a small proportion of carbon, is of low rank.

**hydrocarbon minerals** (*n.pl.*) a general term for compounds of carbon and hydrogen that occur as minerals: coal, bitumens (↑), oil (p.144), nature gas (p.144), and asphalt (↑).

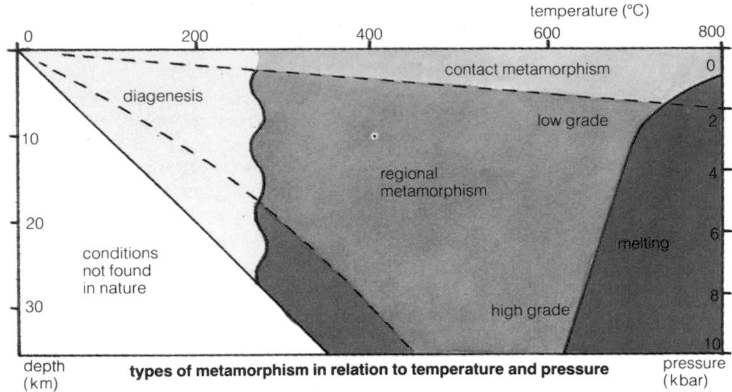


**types of metamorphism in relation to temperature and pressure**

**metamorphism** (*n*) the processes by which rocks are changed by the action of heat or pressure, or both. The rocks changed by metamorphism may be sedimentary (p.80) or igneous (p.62), or even metamorphic. The changes brought about by metamorphism are in the mineral composition, texture (p.72), or structure (p.122) of the rocks. Changes that take place at the Earth's surface (weathering or diagenesis (pp.20, 84)) are not included under metamorphism. **metamorphic** (*adj*), **metamorphose** (*v*).

**unmetamorphosed** (*adj*), not metamorphosed (↑).

**thermal metamorphism** metamorphism (↑) produced by the action of heat on a rock, for example, by an igneous intrusion (p.64).

**contact metamorphism** metamorphism (↑) produced by an igneous intrusion (p.64); largely thermal metamorphism (↑), although some deformation (p.122) may also take place.

**pyrometamorphism** (*n*) an intense form of thermal metamorphism (↑) produced by contact with an igneous intrusion (p.64) at a very high temperature. **pyrometamorphic** (*adj*).

**dynamic metamorphism** metamorphism (↑) produced by mechanical stress (p.122), without heat.

**regional metamorphism** metamorphism (↑) produced by heat and pressure affecting the rocks of a large area (thousands of square kilometres in extent).

**dynamothermal metamorphism** = regional metamorphism (↑).

**metasomatism** (*n*) metamorphism (↑) in which material is added to a rock or taken away from it by liquids or gases passing through the rock. **metasomatic** (*adj*).

**meta-** (*prefix*) meta- at the beginning of a rock name shows that the rock concerned has been metamorphosed (↑), e.g. metaquartzite.

**metamorphic grade** the level that metamorphism (↑) has reached in a rock. It is thus possible to speak, for example, of 'high-grade metamorphism' or of 'metamorphism of medium grade'.

**isograd** (*n*) a line joining points at which the rocks are of the same metamorphic grade (↑).

**prograde** (*adj*) prograde metamorphism changes a rock from a lower to a higher metamorphic grade (↑); it is the normal form of metamorphism.

**retrograde** (*adj*) retrograde metamorphism changes a rock from a higher to a lower metamorphic grade (↑), for example, when a high-grade metamorphic rock is later heated for a long time at a temperature that is lower than the temperature at which the earlier metamorphism took place. *See also* **prograde** (↑).

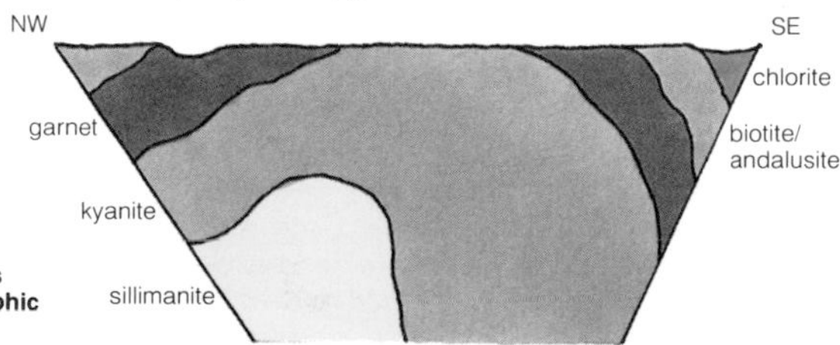


**section across Scottish Highlands showing metamorphic zones**

**metamorphic zones** a metamorphic zone is an area of metamorphic rock that can be mapped in the field by some special character – e.g. the presence of an 'index' mineral (↓).

**index mineral** a mineral produced by metamorphism that is used to characterize a metamorphic zone (↑).

**metamorphic facies** a group of rocks of varying composition, all of which have been metamorphosed under similar conditions.

**greenschist facies** a metamorphic facies (↑) developed at high pressures (from about 2 to 6 kbar) and relatively low temperatures (200 to 500°C). The minerals albite, biotite, muscovite, chlorite, epidote, tremolite, and actinolite (pp.55, 56, 57, 61) are typical of the facies. A characteristic green colour is given to the rocks by the chlorite and epidote.

**amphibolite facies** a metamorphic facies (↑) developed at temperatures of 300 to 600°C and high pressure. Hornblende (p.57) and garnet (p.58) are characteristic minerals of the amphibolite facies.

**granulite facies** a metamorphic facies (↑) developed at temperatures of 700 to 800°C at high pressure and under dry conditions. Characteristic minerals are plagioclase, pyroxene, hornblende, and diopside (pp.55, 57).

**eclogite facies** a metamorphic facies (↑) developed at high pressures (greater than 10 kbar) and high temperatures (600°C or more), characterized by the metamorphic rock eclogite (p.97).

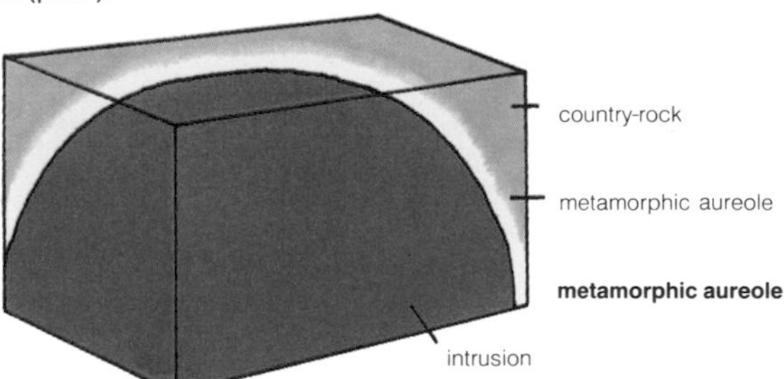


**metamorphic aureole**

**metamorphic aureole** an area round a plutonic igneous intrusion (p.64) in which the country-rock (p.65) has been metamorphosed.

**contact zone** = metamorphic aureole (↑).

**ghost stratigraphy** signs of an earlier stratification (p.80) seen in highly metamorphosed rocks.

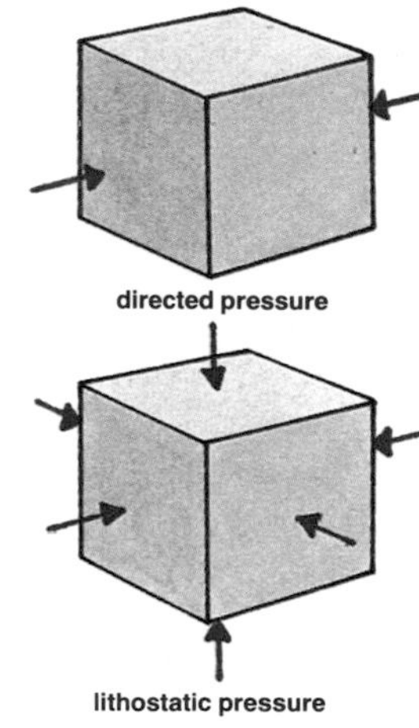
directed pressure

lithostatic pressure

**pressure** (*n*) force per unit area.

**lithostatic pressure** the pressure (force) at depth in the Earth that is due to the weight of rocks above (the *superincumbent load*). Lithostatic pressure increases rapidly with increasing depth.

**hydrostatic pressure** the pressure (↑) at a point in a body of liquid that is due to the weight of liquid above it; *see* **lithostatic pressure** (↑).

**load pressure** = lithostatic pressure (↑).

**confining pressure** = lithostatic pressure (↑).

**directed pressure** a pressure (force) that is applied in such a way that it has a direction.

**pore-fluid pressure** when a rock is heated, minerals containing volatiles (p.18) such as water or carbon dioxide may break down, releasing these volatiles. The pressure that is produced by the volatiles is called the *pore-fluid pressure*. Like lithostatic pressure (↑) it is the same in all directions.

**syntexis** (*n*) a process in which rocks of more than one kind are melted down at great depth to form a new magma (p.62). Country-rock (p.65) may or may not be melted and taken into the mixture that forms the new magma. Compare **anatexis** (↓).

**anatexis** (*n*) the assimilation (p.62) and remelting of rocks by a magma (p.62). Compare **syntexis** (↑).

**rheomorphism** (*n*) the melting of part or the whole of a rock so that it is able to flow and be deformed (p.122).

**recrystallization** (*n*) a process in which the original crystals in a rock are dissolved a little at a time and form a new set of crystals. The new crystals are usually larger than the original ones.

**geothermal heat flow** the rate at which heat passes from the interior of the Earth to the atmosphere.

**geothermal gradient** the rise in temperature that is found with increasing depth below the Earth's surface. The average gradient below the continents is about 30 °C $km^{-1}$.

**isotherm** (*n*) a line joining points that are at the same temperature.

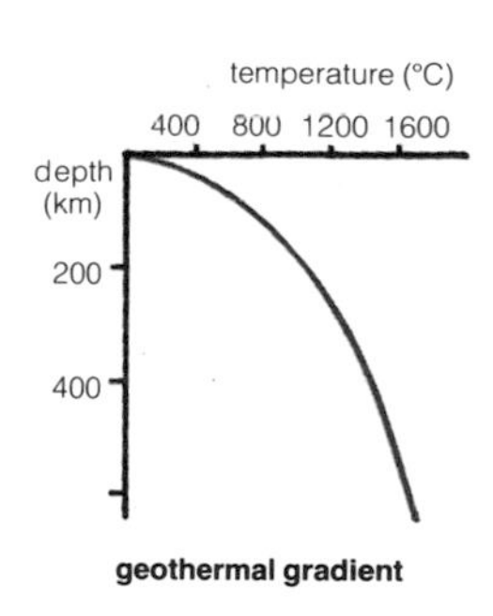

geothermal gradient

**cataclasis** (*n*) the mechanical breaking up of a rock by dynamic metamorphism (p.90). **cataclastic** (*adj*).

**idioblastic** (*adj*) describes a metamorphic texture (p.72) in which the mineral grains show their full crystal form. **idioblast** (*n*).

**crystalloblastic** (*adj*) describes a metamorphic texture produced by the recrystallization (p.93) of the minerals in a rock. **crystalloblast** (*n*).

**porphyroblastic** (*adj*) describes large euhedral (p.45) crystals in a metamorphic rock that are in a groundmass (p.73) of finer grain. **porphyroblast** (*n*).

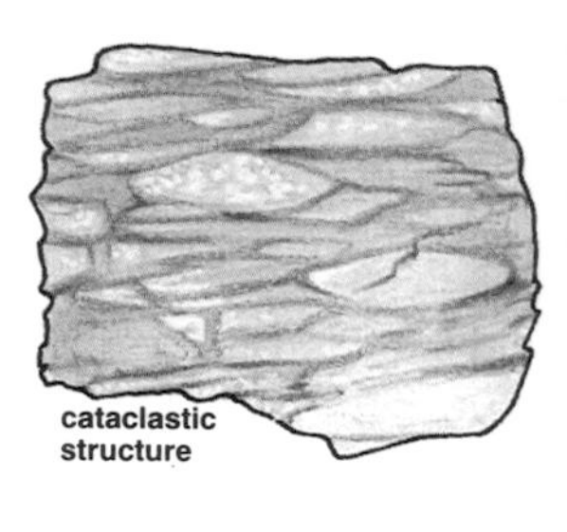
cataclastic structure

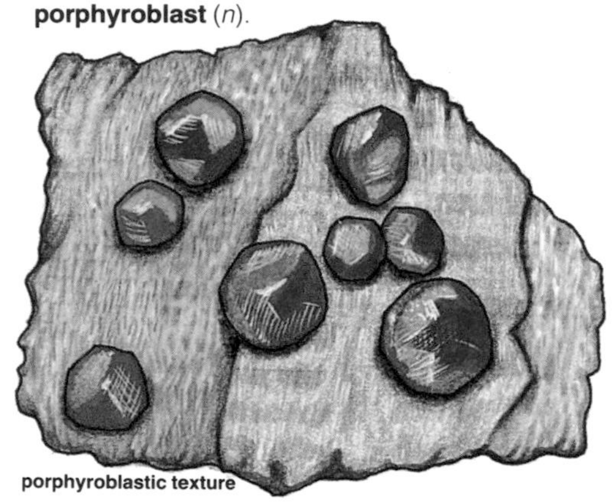
porphyroblastic texture

**xenoblastic** (*adj*) describes a metamorphic rock with mineral grains that do not show crystal faces (p.40). **xenoblast** (*n*).

**relict structure** an original structure that can still be seen in a metamorphic rock after the original minerals have been replaced.

**palimpsest structure** a metamorphic structure showing something of the original texture of the rock. More or less the same as relict structure (↑).

**poikiloblastic** (*adj*) describes a metamorphic texture in which new minerals have formed during recrystallization (p.93) around relics of the original minerals.

**granoblastic** (*adj*) describes an equigranular (p.73) metamorphic texture.

**foliation** (*n*) the arrangement of mineral grains in layers in a rock. **foliated** (*adj*).

**lineation** (*n*) a structure in or on a rock that forms lines.

**schistosity** (*n*) the arrangement of minerals in a coarse-grained metamorphic rock in layers that are parallel or almost parallel.

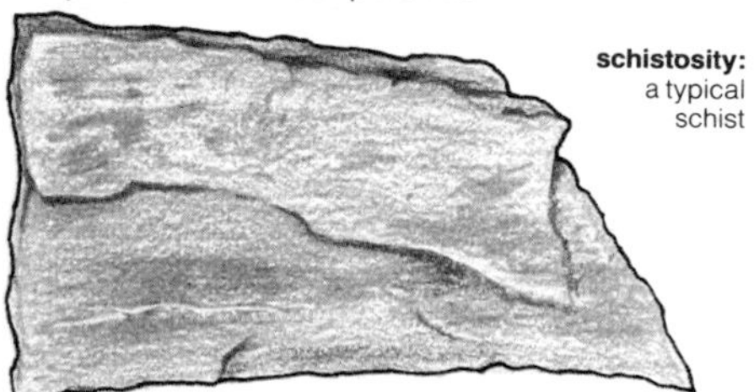

**schistosity:** a typical schist

**cleavage, rock cleavage** if a rock will break or split along smooth planes that are more or less parallel it is said to show cleavage. If it is necessary to draw attention to the difference between this form of cleavage and cleavage as shown by crystals (p.40), the term 'rock cleavage' may be used.

**slaty cleavage** a form of cleavage (↑) that is developed in fine-grained rocks when they are put under great pressure. The cleavage planes are more or less parallel to the axial planes (p.123) of the folds and cut across the bedding planes (p.80).

**flow cleavage** a form of cleavage (↑) in which new minerals grow and the bedding nearly disappears. Further recrystallization (p.93) leads to schistosity (↑).

**fracture cleavage** fine-grained rocks that have been deformed (p.122) may tend to split along closely spaced parallel planes; this is fracture cleavage, also called *shear cleavage* or *false cleavage*.

**shear cleavage** = fracture cleavage (↑).

**false cleavage** = fracture cleavage (↑).

**ptygmatic** (*adj*) a word used to describe free folding or flow folding, e.g. of quartzofeldspathic veins (p.145) in metamorphic or granitized rocks (p.63).

**quartzite** (*n*) the metamorphic equivalent of a quartz sandstone; almost pure silica, $SiO_2$, recrystallized (p.93) into a mass of quartz crystals fitting closely together.

**metaquartzite** (*n*) = quartzite (↑).

**marble** (*n*) a metamorphosed limestone. The calcium carbonate ($CaCO_3$) of the limestone is recrystallized (p.93) as calcite (p.51). The word 'marble' is also used outside geology for various sedimentary and other rocks that can be polished.

marble

**hornfels** (*n*) a hard, fine-grained rock produced by thermal metamorphism (p.90) of sediments. Hornfelses are found at the margins of igneous intrusions (p.64).

**skarn** (*n*) a metamorphic rock produced by the thermal metamorphism (p.90) and metasomatism (p.90) of an impure limestone or dolomite (p.51).

**slate** (*n*) a fine-grained argillaceous rock (p.85) that has a good cleavage (p.44), i.e. it splits easily into thin plates. Slates are produced by dynamic metamorphism (p.90) or regional metamorphism (p.90) of low grade (p.91).

**phyllite** (*n*) a metamorphic rock of higher grade (p.91) than a slate (↑) but of lower grade than a mica-schist (p.55, ↓). Phyllites are produced by regional metamorphism at low temperatures. They are of coarser grain than slates and show better cleavage than mica-schists. They show a characteristic lustre (p.47).

**psammite** (*n*) an arenaceous rock (p.85); usually a metamorphosed arenaceous rock. **psammitic** (*adj*).

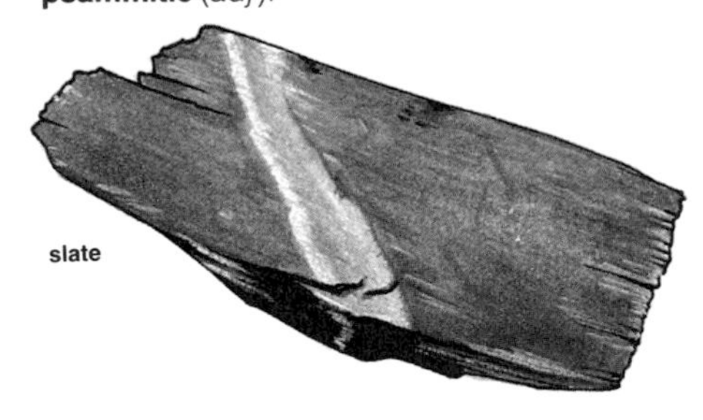
slate

gneiss

**schist** (*n*) a foliated (p.45) metamorphic rock of medium to coarse grain. Schists are produced by regional metamorphism (p.90). The name of a mineral may be added for a particular type of schist, e.g. mica-schist, hornblende-schist.

**gneiss** (*n*) a foliated (p.45) metamorphic rock of coarse grain (p.72) with a banded appearance. Gneisses are produced by regional metamorphism (p.90) of high grade (p.91). **gneissose** (*adj*).

**orthogneiss** (*n*) a gneiss (↑) formed by the metamorphism of an igneous rock (p.62).

**paragneiss** (*n*) a gneiss (↑) formed by the metamorphism of a sedimentary rock (p.80).

**augen gneiss** a gneiss (↑) in which the quartz-feldspar bands are in places thickened and of coarse grain (p.72), forming masses shaped like eyes.

**eclogite** (*n*) a metamorphic rock composed mainly of red garnet (p.58) and green pyroxene (p.57); usually of coarse grain (p.72). The chemical composition of eclogite is similar to that of a basalt or gabbro (p.76). Eclogites are formed at high temperature and pressure. **eclogitic** (*adj*).

**migmatite** (*n*) a mixed rock formed by the injection (p.64) of granitic material between sheets of an existing metamorphic rock. Migmatites are generally formed under regional metamorphism (p.90) of high grade (p.91). **migmatitic** (*adj*).

**tectonite** (*n*) a metamorphic rock in which the minerals tend to point in the same direction, i.e. have a *preferred orientation* or *deformational fabric*. The preferred orientation is the result of crystallization under stress (p.122).

**mylonite** (*n*) a metamorphic rock produced by pressure and rubbing, or *cataclasis* (p.94). The rocks are torn and rolled out to form a fine-grained rock that may show some foliation (p.95). **mylonitic** (*adj*).

**granulite** (*n*) a metamorphic rock with a granular texture (p.72), usually composed of quartz, feldspars, garnets, and pyroxenes (pp.55–8). Granulites are produced by regional metamorphism (p.90). *See also* **granulite facies** (p.92). **granulitic** (*adj*).

**palaeontology** (*n*) the study of the life of past geological times; the study of fossils (↓). **palaeontological** (*adj*).
**paleontology** (*n*) American spelling of palaeontology (↑). **paleontologic** (*adj*).
**biosphere** (*n*) that part of the world in which living things are present: the surface of the land, the soil, the seas, and the air.
**organism** (*n*) a living individual.
**fossil** (*n*) the remains of an animal or plant preserved in a rock; a cast (↓) or impression or a trace (↓) of an animal or plant in a rock. **fossil, fossilized** (*adj*), e.g. a fossil fish, fossilized wood; **fossilize** (*v*).
**fossiliferous** (*adj*) containing fossils (↑).
**fossil record** all the remains of past animal and plant life found in the rocks.
**mould** (*n*) the impression left in a rock by a fossil (or other object). A mould may be *external* (an impression made by the outside of the fossil) or *internal* (a cast (↓) of the inside of a fossil).

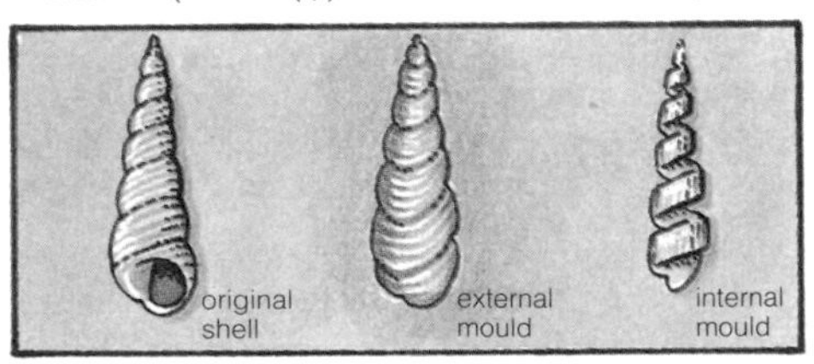


**cast** (*n*) a copy of a fossil (or other object) formed by the filling in of a mould (↑). A cast may be *internal* or *external*, i.e. of the inside or of the outside of the fossil.
**trace fossil** a sedimentary structure (p.83) formed by an animal moving across or moving in the sediment when it was being deposited, e.g. tracks, footprints, and burrows (holes made by animals).

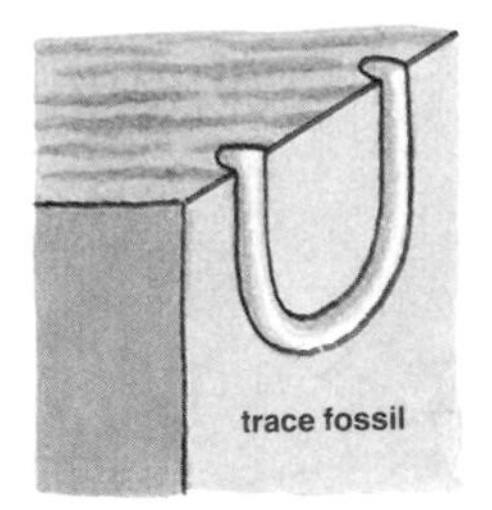


**fauna** (*n*) the animals that live together in any one place or area at a particular time. A *fossil fauna* consists of all the animals that are found as fossils in a particular stratigraphical unit (p.113) – a bed, for example – in a particular area. **faunal** (*adj*). *See also* **flora** (p.110).

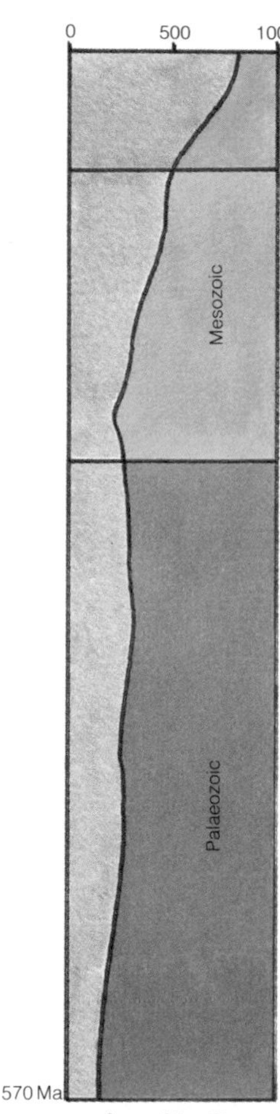


**numbers of families of all groups during Phanerozoic time**

**microfossil** (*n*) a very small fossil (↑): one that can be seen only with a microscope. Some microfossils are important in stratigraphical palaeontology (p.111), e.g. the Foraminifera (p.104).

**micropalaeonotology** (*n*) the study of microfossils (↑). **micropalaeontological** (*adj*).

**microfauna** (*n*) a fauna (↑) of microfossils (↑).

**taphonomy** (*n*) the study of the ways in which fossils (↑) are formed.

**taxonomy** (*n*) the science of arranging animals and plants, whether living or fossilized (↑), in groups or classes according to their structure. **taxonomic** (*adj*).

**taxon** (*n*) (*taxa*) a taxonomic group or unit of classification, e.g. a genus (↓), a family (↓). **taxonomic** (*adj*).

**phylum** (*n*) (*phyla*) one of the main divisions of the animal or plant kingdom.

**class** (*n*) a division of a phylum (↑). e.g. Insecta (p.108), Mammalia (p.109).

**order** (*n*) a division of a class (↑).

**superfamily** (*n*) a division of an order (↑) containing two or more families.

**family** (*n*) a division of an order (↑). Names of families end in -idae.

**subfamily** (*n*) a division of a family (↑). Names of subfamilies end in -inae.

**genus** (*n*) (*genera*) a division of a family (↑), containing one or more species (↓). The name of a genus is a Latin word, written with a capital letter, e.g. *Lingula*.

**species** (*n*) a division of a genus (↑). The members of a species are all very much alike. In living forms they *interbreed* among themselves, i.e. they can become parents of young, who can in turn also become parents. Pairs from different species cannot, on the other hand, produce young. The name of a fossil species is a Latin word, written with a small letter, which comes after the name of the genus, e.g. *Didymograptus murchisoni*.

**type** (*n*) a fossil that represents the characters of a species (↑), a *holotype*; or a genus (↑), a *genotype*.

**palaeoecology** (*n*) the study of fossil animals and plants in relation to the conditions under which they lived – the environment (p.81).
**palaeobiogeography** (*n*) the study of the way in which animals and plants were arranged in space on the surface of the Earth in the geological past.
**habitat** (*n*) the environment (p.81) in which an animal or plant lives or lived.
**assemblage** (*n*) (1) all the fossils that are present in a particular bed or stratum (p.80); (2) the fossils of a species (p.99), or some other small group, from a particular horizon (p.112) or place; (3) a group of fossils found by themselves that are thought to belong to one animal.
**fossil community** a group of fossils found in the same place where they lived together.

**a shell-bed community with lamellibranchs, ammonites, echinoids, etc.**

**plankton** (*n*) all the organisms that float in the sea or in a lake and are carried about by the movement of the water; e.g. Foraminifera (p.104) and Radiolaria (p.104). **planktonic** (*adj*).
**microplankton** (*n*) the smallest members of the plankton (↑); those that cannot easily be seen with the unaided eye.
**phytoplankton** (*n*) all the plant forms that float in the sea or in lakes. *See also* **plankton** (↑).
**nekton** (*n*) all the animals that swim in the sea or in lakes.
**benthos** (*n*) all the animals and plants that live on the sea floor. **benthonic** (*adj*).
**pelagic** (*adj*) animals that live in the open sea but not on the sea floor are called pelagic. They include the nekton (↑) and the plankton (↑).

**a sessile fossil: a crinoid**

**aerobic** (*adj*) needing free oxygen in order to live or be active. *See also* **anaerobic** (↓).
**anaerobic** (*adj*) not needing oxygen in order to live or be active. *See also* **aerobic** (↑).
**epifauna** (*n*) a fauna (p.98) that in life is fixed to another and larger organism (p.98).
**epizoon** (*n*) an organism of an epifauna (↑). **epizoan** (*adj*).
**sessile** (*adj*) describes an organism (p.98) that is closely attached to a surface, such as the sea floor or another organism. Applied to benthos (↑), 'sessile' means attached to the sea floor. *See also* **sedentary** (↓).
**sedentary** (*adj*) describes an organism (p.98) that is attached, as, for example, an oyster. *See also* **sessile** (↑).
**biocoenose** (*n*) an assemblage (↑) of organisms (p.98) that live together as one community (↑).
**biolith** (*n*) a deposit of organic (p.17) material or material formed by the activities of organisms.

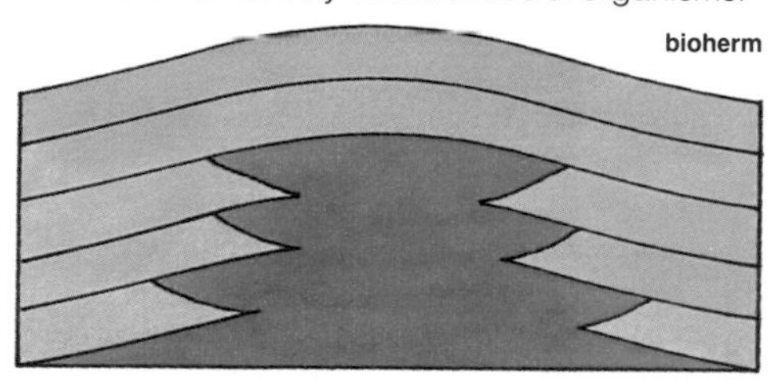
**bioherm**

**bioherm** (*n*) an organic deposit (p.80) built largely or entirely of the remains of fixed organisms (p.98); a fossil reef (p.38). A bioherm is usually shaped like a small hill. It is a special type of biolith (↑).
**biostrome** (*n*) a mass of organic material in the form of a sheet or bed (p.80) built by sedentary organisms (↑, p.98) that have been preserved in place. *See also* **bioherm** (↑).
**biogenic** (*adj*) produced by organisms (p.98).
**stromatolites** (*n*) rounded sedimentary structures (p.83) formed by the plants called algae (p.110), which live in water. The oldest stromatolites are of Pre-Cambrian age (p.114) and are among the oldest fossils known.

**evolution** (*n*) the process by which new forms of living things can develop from earlier forms by passing on small changes from one generation to the next. **evolutionary** (*adj*), **evolve** (*v*).

**adaptation** (*n*) a character of an organism (p.98) that fits it for a particular environment (p.81); (2) the process by which an organism (p.98) is changed to become more fit for its environment (p.81). **adaptive** (*adj*), **adapt** (*v*).

**natural selection** the process by which the weaker and the less well-adapted (↑) individuals in a population – the *unfit* – tend to be taken out of the population without producing young to continue a form that is less valuable.

**adaptive radiation** the development of new species (p.99) that takes place when the descendants of a taxon (p.99) evolve (↑) by natural selection (↑) in fitting themselves to various environments (p.81).

**mutation** (*n*) a discontinuous change in a gene (p.156) or an organism (p.98) that can be inherited, i.e. passed on to its descendants. **mutate** (*v*).

**ancestral** (*adj*) referring to organisms (p.98) from which later organisms are descended.

**ontogeny** (*n*) the course followed by the life history of an individual animal or plant.

**phylogeny** (*n*) the course followed by the evolution (↑) of a species (p.99) or other taxonomic (p.99) group. **phylogenetic** (*adj*).

**diversity (species)** the range of variation that is shown by a species (p.99).

**diversification** (*n*) the process of becoming more diverse, that is, more different. **diversify** (*v*).

**lineage** (*n*) a line of descent from earlier members of the same or a similar group of animals or plants; a series of fossils that show a course of evolution (↑).

**divergence** (*n*) the development of a new population of organisms from an earlier one. **diverge** (*v*), **divergent** (*adj*).

**radiation** (*n*) the evolutionary divergence (↑) of a group of species. **radiate** (*v*), **radiating** (*adj*).

**extinction** (*n*) the dying out of a (whole) group of animals or plants. **extinct** (*adj*).

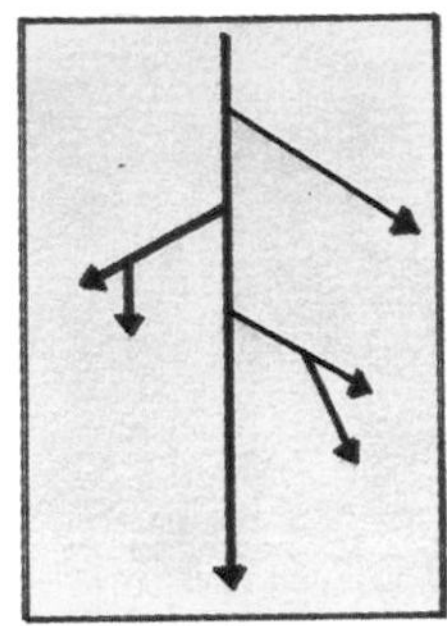

**phylogenetic tree**

**extinctions and appearances of new forms during**

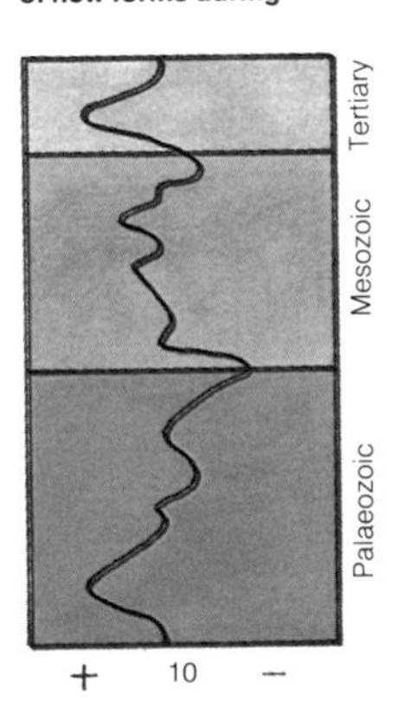

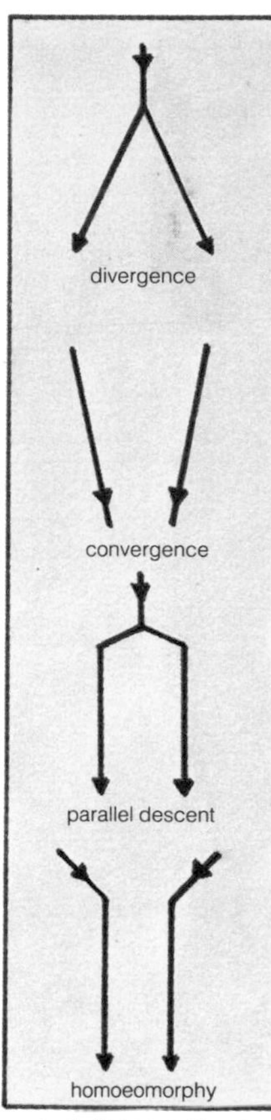


**evolutionary patterns**

**convergence** (*n*) the development of similar forms in different groups of plants or animals at different places or at different times because of the effects produced by similar environments on their separate evolutions (↑). **convergent** (*adj*).

**polyphyletic** (*adj*) a group of organisms (p.98) is polyphyletic if its members have evolved (↑) from different series of earlier forms by convergent evolution (↑).

**trend** (*n*) the evolution (↑) of a particular structural feature within a group of organisms (p.98).

**transient** (*n*) a stage in the phylogeny (↑) of a species; a stage in any closely spaced evolutionary (↑) series (p.159).

**homoeomorphy** (*n*) the occurrence of similar forms (shapes) in members of the same phylum (p.99).

**explosive evolution** a diversification (↑) that for a time takes place much more rapidly than at other times; e.g. the very rapid evolution of the fishes in the late Silurian and early Devonian (p.114).

**evolutionary burst** = explosive evolution (↑).

**quantum evolution** the sudden appearance within a short space of geological time of large taxonomic (p.99) units, e.g. orders (p.99). *See also* **explosive evolution** (↑).

**speciation** (*n*) the production of new species (p.99) by the splitting or division of earlier species in the course of evolution (↑).

**bioseries** (*n*) an evolutionary (↑) series of fossils. It may be a series (p.159) of whole individuals or a series of specimens that show trends in particular features.

**cladogenesis** (*n*) the development of species (p.99) by division of the line of descent. *See diagram*.

**cladistics** (*n*) a cladistic classification is one that is based on the branching pattern (the *cladogram*) of the evolution of a group of animals. Groups that have separated from each other more recently are put closer together than those that have separated at earlier times. This type of classification is unlike those that have been used in the past.

**Invertebrata** (*n.pl.*) the invertebrates: all animals without backbones (the long row of bones in the middle of the back). **invertebrate** (*adj*).

**Protozoa** (*n.pl.*) the phylum (p.99) of the simplest animals, consisting of only one cell. Most of them are very small. The Protozoa include the Foraminifera (↓) and Radiolaria (↓). **protozoan** (*adj*).

**Foraminifera** (*n.pl.*) Protozoa (↑) with shells, which are commonly divided into parts (*chambers*). The shell may be made of calcite (p.51) or arenaceous material (p.85). Most Foraminifera are marine. The Foraminifera are valuable in stratigraphical correlation (p.117), especially in the Tertiary (p.115).

**forams** (*n.pl.*) short for Foraminifera (↑).

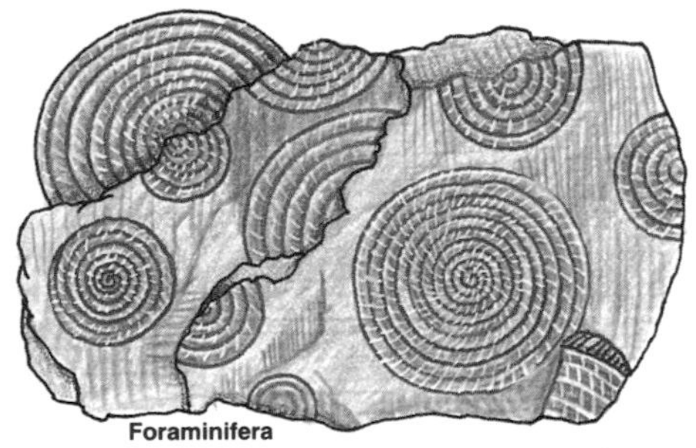

**Foraminifera**

**Radiolaria** (*n.pl.*) marine planktonic (p.100) Protozoa (↑). They have skeletons of silica (p.16). Radiolaria are not common as fossils but are known from the Pre-Cambrian (p.114) onwards.

**Metazoa** (*n.pl.*) animals with many cells (p.153). **metazoan** (*adj*).

**Porifera** (*n.pl.*) the phylum (p.99) of the *sponges*, the simplest of the Metazoa (↑). They live in water, most of them in the sea. Small pieces of the internal skeletons of sponges occur as fossils (sponge spicules).

**Archaeocyatha** (*n.pl.*) marine animals of a simple kind with a calcareous (p.86) skeleton, usually shaped like a cone. They lived on the bottom of the sea but are now extinct.

**Coelentera** (*n.pl.*) a phylum of animals with radial symmetry (p.42): symmetry like that of a wheel. They have a mouth and a space inside the body (the *coelenteron*) in which they digest their food. The Coelentera include the classes Scyphozoa (jellyfish), Hydrozoa (hydra and obelia), and Anthozoa (corals and sea-anemones). The corals (↓) are the most important fossil coelenterates.

**coral**

**corals** (*n.pl.*) marine animals living on the sea floor with skeletons made of calcium carbonate, $CaCO_3$. They belong to the class Anthozoa: see Coelentera (↑). As fossils the corals are important in two ways: as builders of coral reefs (rising to just below or just above the surface of the water) and as zone fossils (p.111).

**Vermes** (*n.pl.*) worms. Animals with soft bodies, found only rarely as fossils.

**worms** *see* **Vermes** (↑).

**Annelida** (*n.pl.*) the annelids, segmented worms (↑), i.e. worms whose bodies are made up of a number of parts or *segments*. Most of them live in the sea. They occur as fossils.

**Brachiopoda** (*n.pl.*) the brachiopods, a phylum (p.99) of marine animals with a shell in two parts, called *valves*, one upper and one lower. A typical brachiopod has a muscle, called the *pedicle*, which passes through a hole at one end of the shell and is fixed to the sea bed. There are two classes (p.99) of Brachiopoda: the Inarticulata and the Articulata. In the Articulata the two halves of the shell are joined by a hinge; the Inarticulata have no hinge. Brachiopods are important as fossils.

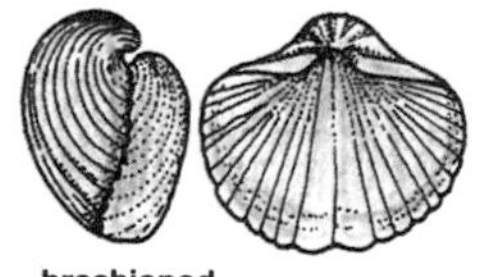

**brachiopod**

**Bryozoa** (*n.pl.*) very small marine animals that form groups or colonies. They have skeletons made of calcium carbonate, $CaCO_3$.

**Polyzoa** (*n.pl.*) = Bryozoa (↑).

**Echinodermata** (*n.pl.*) the echinoderms, a phylum (p.99) of marine animals with skeletons made of calcium carbonate, $CaCO_3$. The skeletons are made up of plates and rods. All echinoderms show radial symmetry (p.42), like that of a wheel, usually fivefold. Two classes, the Crinoidea and the Echinoidea (↓), are of geological importance.

**Crinoidea** (*n.pl.*) the crinoids, a class (p.99) of the Echinodermata (↑). The Crinoidea include the modern sea-lilies. Some crinoids are sessile (p.101); others swim freely. The fixed crinoids have a stem like a flower with a cup (called the *calyx*) at the top. Crinoids are important as index fossils (p.111).

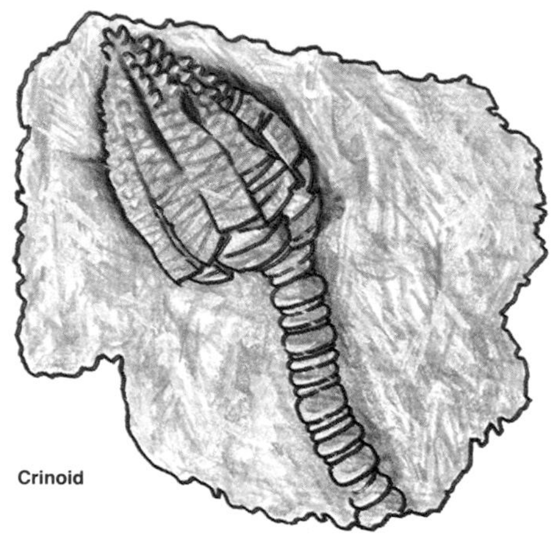

Crinoid

**Echinoidea** (*n.pl.*) the echinoids or sea-urchins. A class of the Echinodermata (↑). The skeleton or *test* is made up of plates of calcium carbonate, $CaCO_3$, and is either round or heart-shaped. It carries rows of sharp *spines*. Echinoids are valuable as zone fossils (p.111) in the Mesozoic and Tertiary (p.115).

Gastropod

Lamellibranch

Ammonite

**Mollusca** (*n.pl.*) the molluscs, a phylum of invertebrates most of which have shells. Some live on land, some in fresh water, and some in the sea. The Mollusca are divided into five classes, of which the Gastropoda (↓), the Lamellibranchiata (↓), and the Cephalopoda (↓) are of geological importance.

**Gastropoda** (*n.pl.*) the gastropods, a class of the Mollusca (↑). They have shells in one piece which are usually twisted like a screw.

**Lamellibranchiata** (*n.pl.*) the lamellibranchs, a class of the Mollusca (↑). Lamellibranchs have a shell of calcium carbonate, $CaCO_3$, which is in two parts or *valves*, one on each side of the animal. The valves are joined by a hinge and both valves have 'teeth' at the hinge. Lamellibranchs are important as fossils.

**Pelecypoda, pelecypods** (*n.pl.*) = Lamellibranchiata (↑).

**Cephalopoda** (*n.pl.*) the cephalopods, a class of the Mollusca (↑). The Cephalopoda have a shell that is divided into separate parts (called *chambers*) by walls (called *septa*; sing. *septum*). All cephalopods are marine.

**Nautiloidea** (*n.pl.*) the nautiloids, a subclass of the Cephalopoda (↑). They have shells that are shaped like a long cone, which may be straight or curved. The septa (↑ Cephalopoda) that divide up the shell are gently curved where they meet the shell and there is a tube (called the *siphuncle*) that runs through the shell, passing through the centres of the septa.

**Ammonoidea** (*n.pl.*) the ammonites, an extinct subclass of the Cephalopoda (↑). The septa (↑ Cephalopoda) that divide up the shell are less simple in shape than those of the Nautiloidea (↑) and the siphuncle (↑) is at the outer edge of the shell, unlike that of a nautiloid. The ammonites are important as fossils.

**Belemnoidea** (*n.pl.*) an extinct group of the Cephalopoda (↑). The most important members geologically are the belemnites. Their fossil shells are in the shape of a long cone with a conical hole at the broad end.

**belemnites** (*n.pl.*) *see* **Belemnoidea** (↑).

**Arthropoda** (*n.pl.*) the arthropods, a large phylum (p.99) of invertebrates (p.104): animals with bodies divided into a number of parts called *segments* and with an external skeleton or *carapace* made of chitin (a hard compound of carbon, hydrogen, nitrogen, and oxygen) and with jointed limbs. The classes (p.99) that are of geological importance are the Trilobita (↓), the Eurypterida (↓), the Ostracoda (↓), and the Crustacea (↓). The arthropods also include the Insecta (insects), the Myriapoda (centipedes and millipedes), and the Arachnida (spiders, scorpions, and mites).

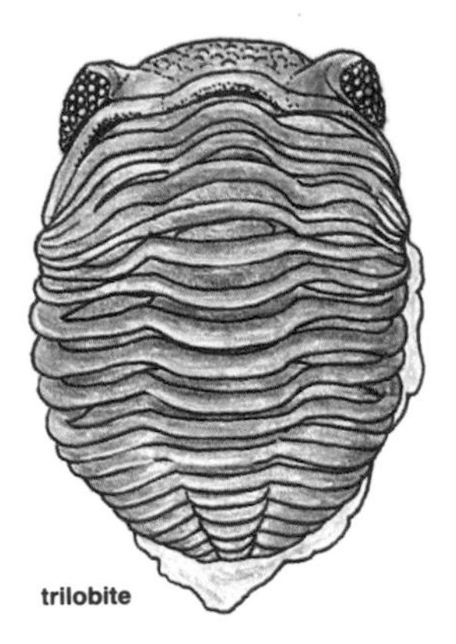
trilobite

**Trilobita** (*n.pl.*) the trilobites, an extinct (p.102) class (p.99) of the Arthropoda (↑). The part of the external skeleton covering the back of the trilobite is divided into three parts. The body of the animal is also divided into a head (called the *cephalon*), a thorax and a tail (the *pygidium*). Trilobites are important in the Palaeozoic (p.114).

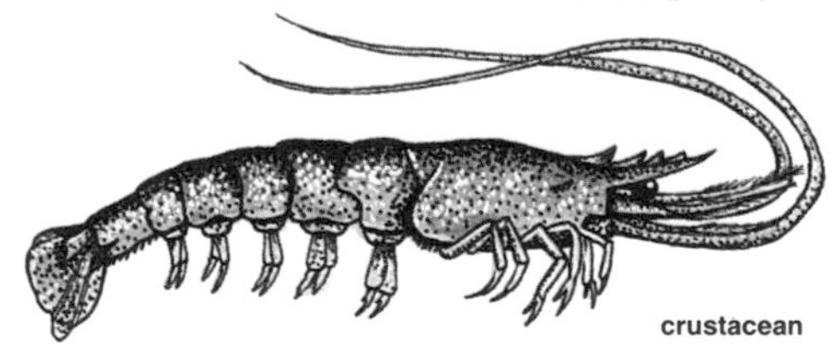
crustacean

**Eurypterida** (*n.pl.*) the eurypterids, an extinct class (p.99) of the Arthropoda (↑). They lived in fresh water and some were up to 2 m long.

**Ostracoda** (*n.pl.*) the ostracods, small arthropods (↑) living in fresh and salt water. The animal lives inside two *valves* (halves) of the *carapace* (shell). Ostracods are used as zone fossils (p.111), for example in the Jurassic (p.115).

**Crustacea** (*n.pl.*) the crustaceans, a class of the Arthropoda (↑). Crabs and lobsters are modern examples. Most crustaceans are marine.

**Arachnida** (*n.pl.*) the arachnids, a division of the Arthropoda (↑) that includes the modern spiders, scorpions, and mites.

**Insecta** (*n.pl.*) the insects, a class of the Arthropoda (↑). Fossils are uncommon.

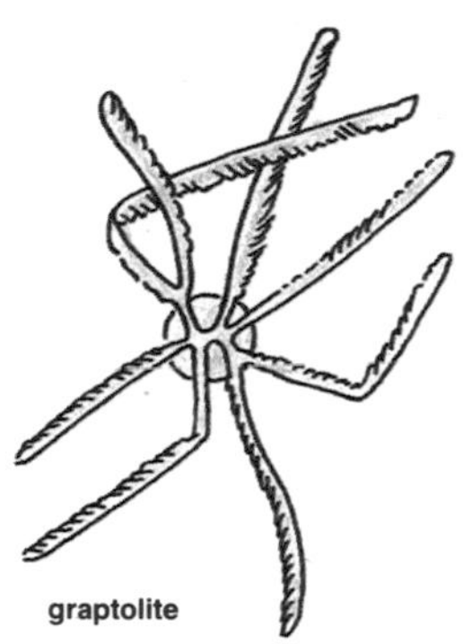
**graptolite**

**Chordata** (*n.pl.*) the chordates, a phylum (p.99) of animals that have at some time in their life-history a hollow tube that can be bent (the *notochord*) or a series of jointed pieces of bone or other material (*vertebrae*) inside the body. Two groups are of geological importance: the graptolites (↓) and the vertebrates (↓).

**Graptolithina** (*n.pl.*) the graptolites. Extinct branching organisms that lived closely together. They are found in rocks of Lower Ordovician to Lower Devonian ages (p.114).

**Vertebrata** (*n.pl.*) vertebrates, the subphylum (p.99) of animals with a skeleton of *vertebrae* inside the body. This skeleton may be made of cartilage (p.152) or of bone. The Vertebrata can be divided into two superclasses, the Agnatha (↓) and the Gnathostoma (↓).

**Pisces** (*n.pl.*) the fishes. They are the earliest and simplest vertebrates (↑). The earliest fossil fish are found in Ordovician rocks (p.114).

**Agnatha** (*n.pl.*) a superclass of the Vertebrata (↑): jawless fish. Fossil forms are found in rocks of Devonian age (p.114).

**Gnathostoma** (*n.pl.*) a superclass of the Vertebrata (↑): vertebrates with jaws. Examples are known from the Devonian (p.114).

**Amphibia** (*n.pl.*) the amphibians, a class of vertebrates (↑) living on land and in water. Fossil amphibia are found in rocks of Upper Devonian to Recent age (p.115). **amphibian** (*adj*).

**Reptilia** (*n.pl.*) the reptiles, a class of vertebrates (↑) which evolved (p.102) from the Amphibia (↑) in Upper Carboniferous times (p.114). **reptilian** (*adj*).

**dinosaurs** (*npl.*) a group of Mesozoic reptiles (↑). Some were very large: up to 35 m long.

**Aves** (*n.pl.*) the birds: a class of the Vertebrata (↑). Very few fossils are known. The earliest are of Upper Jurassic age (p.115). **avian** (*adj*).

**Mammalia** (*n.pl.*) the mammals, a class of warm-blooded vertebrates with hair and teeth of various shapes. The mother feeds the young with her milk. Fossil mammals range from Jurassic to Recent (p.115) and are most important in the Tertiary (p.115). **mammalian** (*adj*).

**palaeobotany** (*n*) the study of fossil plants. **palaeobotanical** (*adj*).

**flora** (*n.pl.*) the plants of a particular place or time. In palaeobotany (↑), the plants in a stratigraphical (p.112) unit or a geological area.

**algae** (*n.pl.*) the sea weeds and related plants. Fossil algae are known from the Pre-Cambrian (p.114).

**Diatomaceae** (*n.pl.*) the diatoms; single-celled algae (↑) whose cell walls are full of silica, $SiO_2$.

**vascular plants** plants having a *vascular system*: a system of cells and tissues (p.160) for carrying water, mineral salts, and food through the plant. The vascular system also gives strength and support to the plant. The vascular plants are divided into two groups: the Pteridophyta, or spore-bearing plants, and the Spermatophyta, or seed-bearing plants (↓).

**Pteridophyta** (*n.pl.*) the spore-bearing plants, one of the main divisions of the plant kingdom. It includes the modern ferns (Filicales), horsetails (Equisitales), and club-mosses (Lycopodiales).

**Pteridospermae** (*n.pl.*) the seed-ferns: an extinct group of plants that were shaped like ferns, a group of flowerless plants but that bore seeds. They were important in the Carboniferous and Permian periods (p.114).

**Spermatophyta** (*n.pl.*) the spermatophytes or seed-bearing plants. They can be divided into the Gymnospermae (↓) or gymnosperms – the conifers etc. – and the Angiospermae (↓) or angiosperms – the flowering plants.

**Gymnospermae** (*n.pl.*) the gymnosperms, one of the main divisions of the plant kingdom. It includes the conifers (Coniferales), ginkgos (Ginkgoales), and the cycads (Cycadales).

**Angiospermae** (*n.pl.*) the angiosperms or flowering plants. One of the main divisions of the plant kingdom and the most important group since the Cretaceous (p.115).

**palynology** (*n*) the study of fossil spores and pollen (the cells and powder by which plants reproduce themselves). Study of fossil spores and pollen is useful in stratigraphical correlation (p.117).

**fossil plants**

**stratigraphical palaeontology** the study of fossils in order to understand the geographical distribution of animals and plants in the geological past and the history of life through geological time.

**zone fossil** a fossil species (p.99) that is chosen as characteristic of those that are present in a particular stratum (p.80). The name of the zone fossil is given to the zone (p.117). A zone fossil should be found only in the stratum that is named after it.

**index fossil** = zone fossil (↑).

**range** (*n*) the range of a fossil is the distance in time covered by its occurrence in the rocks, from its first appearance to its last.

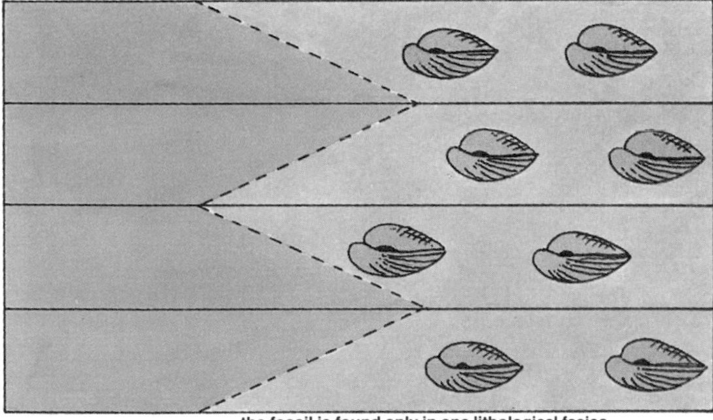

the fossil is found only in one lithological facies

**facies fossil, facies fauna** a facies fossil or a facies fauna is one that occurs only in rocks of a particular rock type or facies (p.117). Such fossils are of little or no use for stratigraphical correlation (p.117) and it is important to know that they are facies fossils.

**derived fossil** a fossil that has been eroded (p.20) from rocks that were deposited earlier and has been deposited again in a younger bed (p.80). A derived fossil is therefore not of the same age as the rocks in which it is found.

**stratigraphy** (*n*) the study of stratified rocks (p.80), their nature, their occurrence, their relationships to each other and their classification. **stratigraphical, stratigraphic** (*USA*) (*adj*).

**historical geology** the study of the history of the Earth. It includes stratigraphy (↑).

**Uniformitarianism** (*n*) the view that geological processes were of the same kind in the past as they are today and produced similar results. *See also* **Catastrophism** (↓).

**Catastrophism** (*n*) the view, no longer held in geology, that the history of the Earth has to be explained by a series of violent events or catastrophes. *See also* **Uniformitarianism** (↑).

**succession** (*n*) the order in which rock-groups appear. When a succession is set out in the form of a table the beds (p.80) are shown with the oldest at the bottom and the youngest at the top.

**superposition** (*n*) the order in which rocks are placed one above the other. The *principle* or *law of superposition* is that in a layered succession (↑) of rocks the lower beds (p.80) will be the older and the upper beds will be the younger (unless the rocks have been turned upside down).

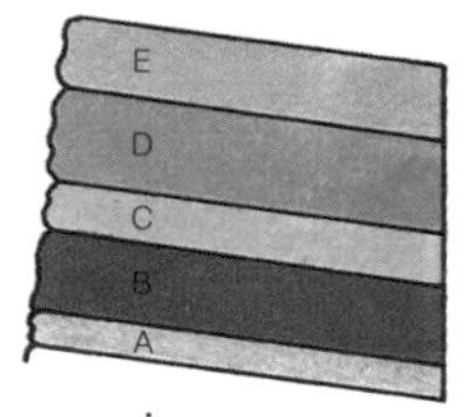


**succession**

bed A is the oldest
bed E the youngest
(if not inverted)

**time plane** a surface within a series of sedimentary rocks that marks a particular moment in geological time.

**horizon** (*n*) (1) a plane of stratification (p.80) that was once horizontal and continuous; (2) a time plane (↑) within a sedimentary series (↓) or a bed (p.80) (usually a thin bed) that contains characteristic fossils or has a characteristic lithology (p.85).

**sequence** (*n*) a succession (↑) of bedded rocks; the stratigraphical (↑) order in which beds appear.

**cyclic sequence** a sequence (↑) of sediments (p.80) repeated in a particular order, e.g. sandstone – shale – limestone. A cyclic sequence is commonly the result of marine transgression (p.119) and regression (p.119).

**cyclothem** (*n*) a unit of a cyclic sequence (↑).

**rhythmic sequence** a cyclic sequence (↑) on a small scale.

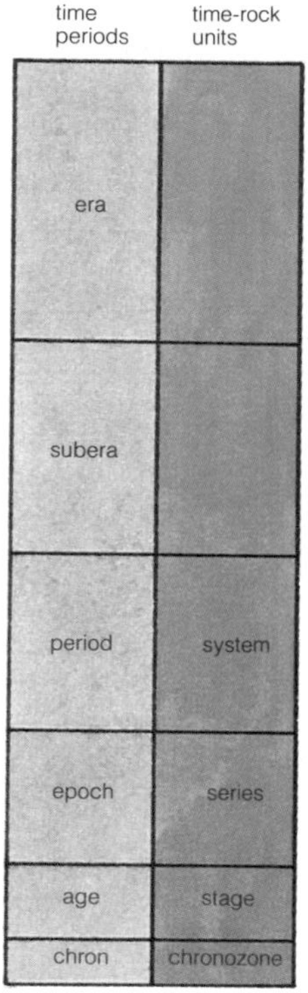

| time periods | time-rock units |
|---|---|
| era | |
| subera | |
| period | system |
| epoch | series |
| age | stage |
| chron | chronozone |

**aeon, eon** (*n*) the largest division of geological time. An aeon is made up of several eras (↓).

**era** (*n*) a division of geological time; made up of several periods (↓) or sub-eras (↓).

**sub-era** (*n*) a division of an era (↑).

**period** (*n*) a large division of geological time; it corresponds to a system (↓).

**epoch** (*n*) a division of geological time; part of a period (↑); it corresponds to a series (↓).

**age** (*n*) a division of geological time; part of an epoch (↑); it corresponds to a stage (↓).

**chron** (*n*) the smallest division of geological time; part of an age.

**chronostratigraphical** (*adj*) a chronostratigraphical unit is a division of the geological column (↓) that is based on geological time. *See also* **lithostratigraphical** (↓), **biostratigraphical** (p.117).

**system** (*n*) one of the major stratigraphical (↑) divisions of the geological column (↑); it corresponds to a geological period (↑).

**series** (*n*) a stratigraphical division within a system (↑); it corresponds to an epoch (↑).

**stage** (*n*) a stratigraphical division within a series (↑); it corresponds to an age (↑).

**lithostratigraphical** (*adj*) a lithostratigraphical unit is one that is based on lithological (p.85) characters rather than on geological time or fossils. *See also* **chronostratigraphical** (↑), **biostratigraphical** (p.117).

**rock-stratigraphical** (*adj*) = lithostratigraphical (↑).

**group** (*n*) a lithostratigraphical term (↑) for a rock unit consisting of two or more formations (↓) that are next to each other in a succession (↑) and are related to each other.

**formation** (*n*) a term for the basic lithostratigraphical (↑) division.

**member** (*n*) a lithostratigraphical (↑) term for a part of a formation (↑).

**bed** (*n*) the smallest lithostratigraphical (↑) division. *See also p.80.*

**geological column** a diagram that shows the divisions of geological time and the succession (↑) for a given area.

**Pre-Cambrian, Precambrian** (*n, adj*) the period of time before the Cambrian (↓), i.e. from the formation of the Earth until about 570 million years ago: about 4000 million years.

**Proterozoic** (*n, adj*) (1) one of two aeons (p.113) into which the Pre-Cambrian (↑) is divided, ranging from 2500 to 570 million years ago. (2) the whole of the Pre-Cambrian.

**Archaean** (*n, adj*) (1) the earlier of two aeons (p.113) into which the Pre-Cambrian (↑) is divided, covering the period from the formation of the Earth until 2500 million years ago; (2) the whole of the Pre-Cambrian.

**basement complex, basement** a general term for igneous (p.62) or metamorphic (p.90) rocks, usually Pre-Cambrian (↑), which cover a wide area and on which rest unmetamorphosed (p.90) sediments of later age.

**Phanerozoic** (*n, adj*) the stratigraphical systems from the Cambrian (↓) to the Recent (↓).

**Palaeozoic** (*n, adj*) the era of geological time that ranges from 570 to 230 million years ago. It is divided into the Lower Palaeozoic, consisting of the Cambrian (↓), Ordovician (↓), and Silurian (↓) periods, and the Upper Palaeozoic, consisting of the Devonian (↓), Carboniferous (↓), and Permian (↓) periods.

**Cambrian** (*n, adj*) the earliest period of the Palaeozoic era, dating from about 570 million years ago to 500 million years ago.

**Ordovician** (*n, adj*) a period of the Palaeozoic Era (↑), dating from 500 to 435 million years ago.

**Silurian** (*n, adj*) a period of the Palaeozoic Era (↑), dating from 435 to 400 million years ago.

**Devonian** (*n, adj*) a period of the Palaeozoic Era (↑), dating from 400 to 345 million years ago.

**Carboniferous** (*n, adj*) a period of the Palaeozoic Era (↑), dating from 345 to 280 million years ago. In the USA it is divided into the Mississippian (below) and the Pennsylvanian (above).

**Mississippian** (*n, adj*) *see* **Carboniferous** (↑).

**Pennsylvanian** (*n, adj*) *see* **Carboniferous** (↑).

**Permian** (*n, adj*) the latest period of the Palaeozoic Era (↑), dating from 280 to 230 million years ago.

Ma
250
300
400
500
Triassic
Permian
Carboniferous
Devonian
Silurian
Ordovician
Cambrian

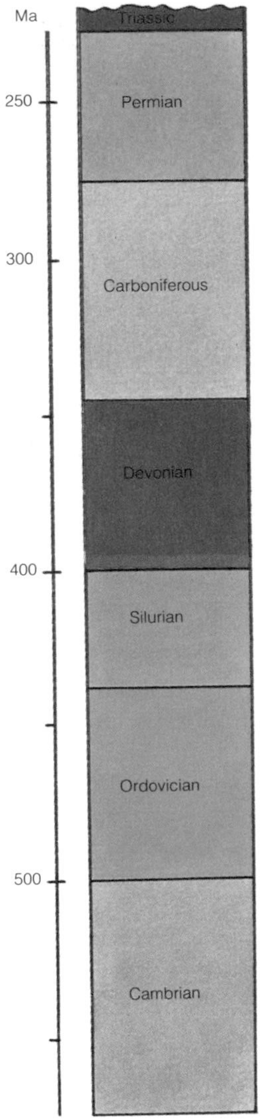

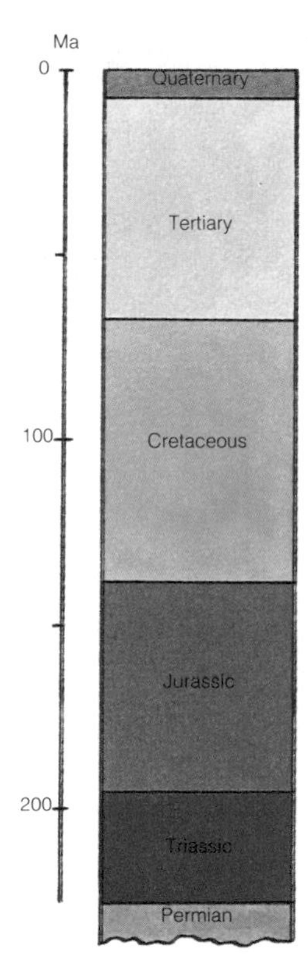


**Mesozoic** (*n, adj*) the era between the Palaeozoic (↑) and the Cainozoic (↓), ranging from 230 to 65 million years ago. It is made up of the Triassic, Jurassic, and Cretaceous periods.

**Triassic** (*n, adj*) the earliest period of the Mesozoic Era (↑), ranging from 230 to 195 million years ago.

**Trias** = Triassic (↑).

**Jurassic** (*n, adj*) one of the three periods of the Mesozoic Era (↑), ranging from 195 to 140 million years ago.

**Cretaceous** (*n, adj*) the youngest of the three periods of the Mesozoic Era (↑), ranging from 140 to 65 million years ago.

**Cainozoic, Cenozoic** (*n, adj*) the era of geological time that follows the Mesozoic (↑), ranging from 65 million years ago to the present. It is made up of the Tertiary and Quaternary sub-eras (↓).

**Tertiary** (*n, adj*) the sub-era between the Cainozoic era (↑) and the Quaternary sub-era (↓), ranging from 65 million years ago to 2 million years ago. It is divided into two periods, the Palaeogene and the Neogene, and five epochs: the Palaeocene, Eocene, Oligocene, Miocene, Pliocene, Pleistocene, and Holocene.

**Palaeocene** (*n, adj*) *see* **Tertiary** (↑).

**Eocene** (*n, adj*) *see* **Tertiary** (↑).

**Oligocene** (*n, adj*) *see* **Tertiary** (↑).

**Miocene** (*n, adj*) *see* **Tertiary** (↑).

**Pliocene** (*n, adj*) *see* **Tertiary** (↑).

**Palaeogene** (*n, adj*) the earlier of the two periods of the Tertiary sub-era (↑). It consists of the Palaeocene, Eocene, and Oligocene epochs (↑).

**Neogene** (*n, adj*) the later of the two periods of the Tertiary sub-era (↑). It consists of the Miocene and Pliocene epochs (↑).

**Quaternary** (*n, adj*) the period from 2 million years ago to the present; a subdivision (sub-era) of the Cainozoic Era (↑). It is divided into two epochs: the Pleistocene (↓) and Holocene (↓).

**Pleistocene** (*n, adj*) an epoch of the Quaternary sub-era (↑); the time of the last ice age.

**Holocene** (*n, adj*) the latest epoch of the Quaternary sub-era (↑); it includes the present time.

**Recent** = Holocene (↑).

**orogenic period** a period of mountain-building. *See also p.132.*

**Caledonian** (*adj*) relating to a period of mountain-building in Ordovician and Devonian times (p.114). The general trend of the Caledonian structures is north-east – south-west.

**Caledonides** (*n*) the former range of mountains that was formed during the Caledonian orogeny (↑), reaching from Norway to Scotland and Ireland.

**Hercynian** (*adj*) relating to the period of mountain-building that took place in late Palaeozoic times (p.114) in Europe.

**Variscan** (*adj*) (1) = Hercynian (↑); (2) relating to a period of mountain-building from the Carboniferous (p.114) to the Triassic (p.115).

**Kimmerian** (*adj*) relating to a period of mountain-building that took place in Jurassic times (p.115) in Europe.

**Alpine** (*adj*) relating to the period of mountain-building in the Tertiary period (p.115) that formed the Alps in Europe.

**Taconic** (*adj*) relating to a period of mountain-building that took place in late Ordovician times (p.114) in North America.

**Acadian** (*adj*) relating to a period of mountain-building that took place in ? late Devonian to end Permian times (p.114) in North America.

**Appalachian** (*adj*) relating to a period of mountain-building that took place in late Palaeozoic times in North America.

**Laramide** (*adj*) relating to a period of mountain-building that took place in late Cretaceous (? Jurassic) to early Eocene times (p.115) in North America.

**synorogenic** (*adj*) taking place at the same time as a period of mountain-building.

**post-orogenic** (*adj*) taking place after a period of mountain-building.

**syntectonic** (*adj*) taking place at the same time as a period of deformation (p.122).

**synkinematic** (*adj*) = syntectonic (↑).

**post-tectonic** (*adj*) taking place after a period of deformation (p.122).

**postkinematic** (*adj*) = post-tectonic (↑).

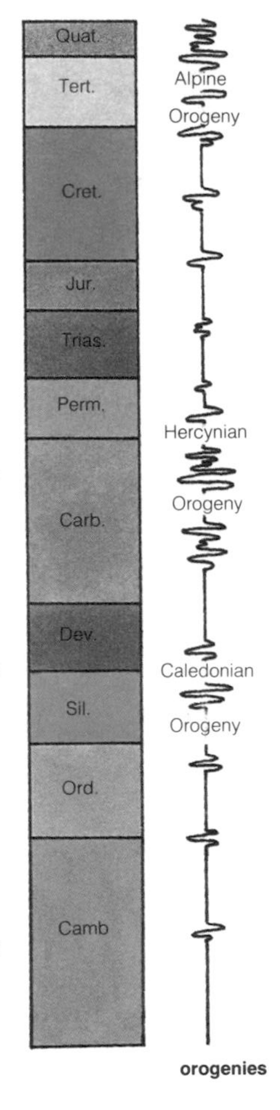


**orogenies**

**biostratigraphical** (*adj*) a biostratigraphical unit is one that is based on fossils rather than on lithological (p.85) characters or on geological time. *See also* **lithostratigraphical** (p.113), **chronostratigraphical** (p.113).

**zone** (*n*) a biostratigraphical (↑) division: a stratigraphical division (p.111) with characteristic fossils. One of the fossils present – the zone fossil – gives the name to the zone. **zonal** (*adj*).

**hemera** (*n*) a small unit of geological time as marked by the rise and fall of a particular species fossil. The word is not now in common use. **hemeras, -ai, -ae** (*pl.*).

**epibole** (*n*) a stratigraphical term for the rocks deposited during a hemera (↑); i.e., the time-rock unit corresponding to a hemera. The word is not now in common use.

**correlation** (*n*) in stratigraphy, the matching of rocks of a particular age that are found in one place with other rocks found elsewhere. Isotopic age-determinations (p.121) provide a means of dating rocks and correlating them; fossils can also be used for correlation. **correlate** (*v*).

**provenance** (*n*) the source area of the materials that form a sedimentary rock; the nature of the rocks from which it has been formed.

**facies** (*n*) (*facies*) the general characters of a sedimentary rock, especially those that indicate the environment (p.81) in which it was deposited.

**lithofacies** (*n*) a facies (↑) that is characterized by a particular rock type.

**biofacies** (*n*) a facies (↑) that is characterized by a particular assemblage (p.100) of fossils.

**diachronous** (*adj*) 'across time'. A word used to describe a bed (p.80) or a stratigraphical unit that is of different ages in different places.

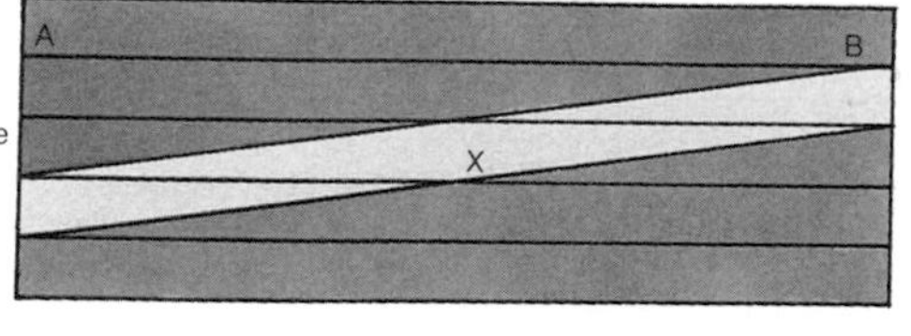


**a diachronous formation**
X is older at A than at B

**type-area, type-locality** a place that is chosen as the example for a stratigraphical unit (p.112).

**outlier** (*n*) an outcrop (p.122) of younger rocks with older rocks all round them, the younger rocks being separated from their main outcrop.

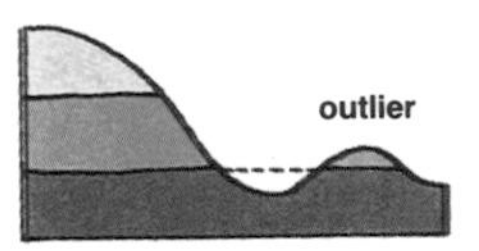


**inlier** (*n*) an outcrop (p.122) of older rocks with younger rocks all round them.

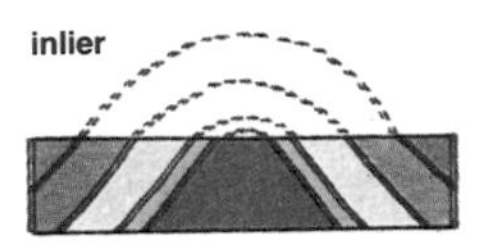


**conformable** (*adj*) beds (p.80) are conformable when they lie on each other in a regular way. **conformity** (*n*).

**unconformity** (*n*) an unconformity is present between sedimentary rocks (p.80) and the rocks on which they rest if a period of non-deposition, i.e. a period during which no sediments were deposited (p.80), has taken place between the formation of the older rocks and the deposition (p.80) of the later sediments. An unconformity may be shown by a difference in dip (p.123) between the two series (*angular unconformity*) or by an irregular surface between them. In some cases the two series may be separated only by a period of non-deposition without later movement. There is then no difference in dip between the older and the newer sediments. This type of unconformity is called a *non-depositional unconformity*, a *non-sequence*, or a *diastem*. **unconformable** (*adj*).

angular unconformity

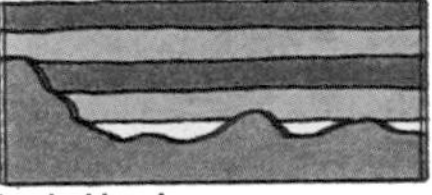
buried landscape

**non-sequence** (*n*) a non-depositional unconformity (↑); an unconformity in which the beds above and below the plane of unconformity (↑) are parallel to each other. The time covered by a non-sequence is usually relatively short.

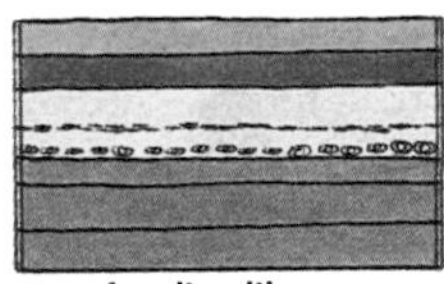
unconformity with basal conglomerate

**diastem** = non-sequence (↑).

**disconformity** (*n*) more or less the same as a non-sequence (↑).

**contemporaneous** (*adj*) occurring at the same time. **contemporaneously** (*adv.*).

**penecontemporaneous** (*adj*) almost at the same time, e.g. the penecontemporaneous erosion of sediments shortly after their deposition.

**intraformational** (*adj*) occurring within one stratigraphical formation (p.113), e.g. an *intraformational conglomerate*, a conglomerate formed by contemporaneous (↑) erosion (p.20) and deposition (p.80).

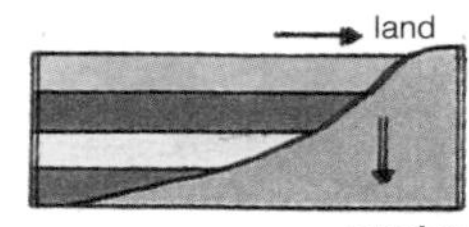

**overlap**

**overstep** (*n*) a relationship that is produced by unconformity (↑) in which the beds at the base of a younger series rest upon older and older beds of the series below the unconformity as they are followed across country. Overstep is the usual relationship produced by a marine transgression (↓). *See also* **overlap** (↓).

**overlap** (*n*) a relationship in which the members of a younger series of sediments spread in turn further and further over an older series below them. Overlap results when a marine transgression (↓) takes place while the surface on which the new rocks are being deposited (p.80) is sinking. *See also* **overstep** (↑).

**offlap** (*n*) a relationship in which younger members of a series of sediments cover a smaller and smaller area as the sequence is followed upwards. Offlap is produced by deposition (p.80) during a marine regression (↓). *See also* **overlap** (↑).

**transgression, marine** (*n*) the spreading of the sea over the land in a relatively short period of geological time. A marine transgression results in overlap (↑). *See also* **regression** (↓).

**regression, marine** (*n*) the opposite of transgression (↑): the movement of the sea away from the land in a relatively short period of geological time. A marine regression produces offlap (↑). *See also* **transgression** (↑).

**interstratified** (*adj*) strata (p.80) laid down or alternating with other strata. *See also* **interbedded** (↓), **intercalated** (↓). **interstratify** (*v*).

**interbedded** (*adj*) deposited (p.80) in sequence (p.112) between one bed and another; used especially of lavas (p.70) between beds in a sedimentary (p.80) succession (p.112).

**intercalated** (*adj*) (1) put into a series of bedded (p.80) rocks after their formation (e.g. a lava); (2) interstratified (↑). *See also* **interbedded** (↑). **intercalate** (*v*).

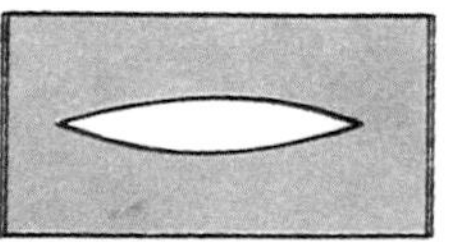
**lens**

**lens** (*n*) a mass of rock or other material that is thick in the centre and thin at the edges.

**lenticular** (*adj*) shaped like a lens (↑).

**geochronology** (*n*) the science of dating rocks and geological events in years. Radiometric dating (↓) and the counting of varves (↓) and tree-rings (↓) are the chief methods used.

**radiometric dating** a method of dating rocks and minerals by measuring the amounts of radioactive elements (p.19) in them and the daughter elements (p.19) into which they decay.

**potassium–argon dating** the isotope (p.19) of potassium $^{40}K$ is radioactive and decays in two ways to yield $^{40}A$ and $^{40}Ca$, isotopes of argon and calcium. In the potassium–argon (K–Ar) method of radiometric dating (↑) the amounts of $^{40}K$ and $^{40}Ar$ are measured in order to calculate the age of the rock. The K–Ar method can be used for ages from 3400 Ma down to 30 000 years.

**K–Ar dating** = potassium–argon dating (↑).

**rubidium–strontium dating** the isotope of rubidium $^{87}Rb$ decays to give $^{87}Sr$, an isotope of strontium. In the rubidium–strontium (Rb–Sr) method of radiometric dating (↑), the amounts of $^{87}Rb$ and $^{87}Sr$ are measured in order to calculate the age of the rock. The Rb–Sr method is used for Pre-Cambrian (p.114) rocks and for igneous and metamorphic rocks (pp.62, 90).

**Rb–Sr dating** = rubidium–strontium dating (↑).

**uranium–lead, lead–lead, and thorium–lead dating** the isotope of uranium $^{238}U$ decays to give $^{206}Pb$; and $^{235}U$, another uranium isotope, decays to give $^{207}Pb$. The ratio of $^{207}Pb$ to $^{206}Pb$ provides a second method of measuring the age of a rock. The thorium–lead (Th–Pb) ratio provides a third method, but the results are less reliable. These three methods are used mainly for rocks and minerals with ages greater than 100 Ma.

**carbon-14 dating** the isotope of carbon $^{14}C$ decays to give $^{14}N$, an isotope of nitrogen with a half-life (p.19) of 5570 years. In the $^{14}C$ method the ratio of these two isotopes is measured in order to find the age of the specimen. The $^{14}C$ method is used for dating events up to 70 000 years before the present.

**$^{14}C$ dating** = carbon-14 dating (↑).

| | | half-life |
|---|---|---|
| $^{40}K$ | $^{40}Ar$ | 1300 Ma |
| $^{87}Rb$ | $^{87}Sr$ | 47000 Ma |
| $^{238}U$ | $^{206}Pb$ | 4510 Ma |
| $^{235}U$ | $^{207}Pb$ | 713 Ma |
| $^{232}Th$ | $^{208}Pb$ | 13900 Ma |
| $^{14}C$ | $^{14}N$ | 5570 years |

**half-life**

**isotopic age** an age (of a rock or mineral) measured by using radiometric methods (↑). The amounts of radioactive isotopes (p.19) are measured in order to calculate the age of the specimen.

**absolute age** an age (of a rock or mineral) expressed in numbers of years. 'Absolute age' is also used to mean 'isotopic age', but this is better avoided.

**apparent age** a radiometric age (↑) that is not the true age of the rock or mineral concerned.

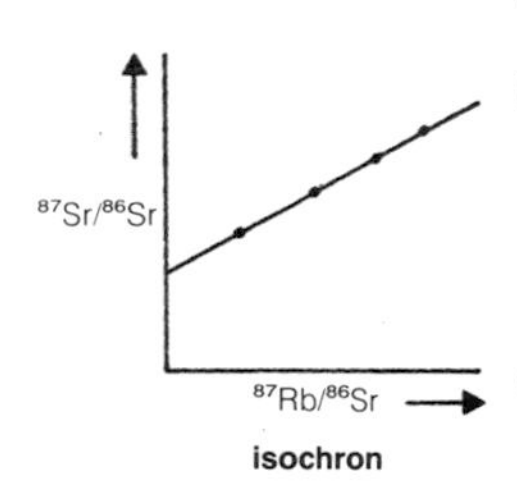


**isochron**

slope of curve gives age

**isochron** (*n*) measurements of isotope ratios (p.19) of several minerals in a rock can be shown on a graph (p.157) in what is known as an *isochron plot*. The isochron is a straight line on the graph and its slope represents the age of the rock.

**concordant** (*adj*) when more than one method of radiometric dating (↑) is used on a mineral or group of minerals in a rock and the dates found agree with each other within the limits of the method they are called *concordant*. *See also* **discordant** (↓).

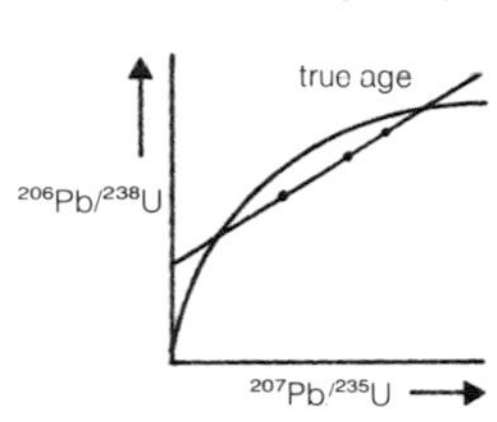


**discordant age-pattern**

**discordant** (*adj*) when more than one method of radiometric dating (↑) is used on a mineral or group of minerals in a rock and the dates found do not agree with each other they are called *discordant*. Discordant ages result when a rock has had a geological history that is not simple. They can provide useful geological information.

**varve-count** (*n*) sediments (p.80) deposited in lakes formed by water flowing from glaciers (p.28) show a layered arrangement because the sediment deposited is fine in winter and coarse in summer. These layers are called *varves*. By counting them the ages of Pleistocene rocks (p.115) can be found.

**tree-ring dating** a tree usually adds a growth-layer to itself every year. By counting the rings in cross-sections of trees and by comparing the patterns of wider and narrower rings in the trees of a region it is possible to measure ages back to 7000 years BP (before the present).

**dendrochronology** (*n*) = tree-ring dating (↑).

**structure** (*n*) the shapes and positions of rock masses and their relationships to each other. **structural** (*adj*).
**tectonics** (*n*) the study of the structure (↑) of the Earth's crust (p.9) or of a particular region. **tectonic** (*adj*).
**outcrop** (*n*) the area where a rock-unit occurs at the surface of the Earth. The rock-unit may be a stratum (p.80), an igneous intrusion (p.64), or any other body. It need not be exposed (↓) at the surface. **crop out** or **outcrop** (*v*).
**exposure** (*n*) a place where rocks can be seen in their natural position and are not covered by vegetation or buildings. **expose** (*v*), **exposed** (*adj*). *See also* **outcrop** (↑).
**isopachyte, isopach** (*n*) a line on a map joining points at which a bed (p.80) has the same thickness.
**stress** (*n*) if a force (p.156) is applied to the surface of a body, such as a mass of rock, the force per unit area is called the *stress*. If the stress acts in a particular direction it is called a *directed stress*; if it acts equally in all directions it is called a *hydrostatic stress*. Stress may be of three types: *compressional* or *compressive*, when the forces are acting towards the centre of the body; *tensile*, when the forces tend to pull the body out; and *shear*, when two forces are acting that tend to turn the body round.

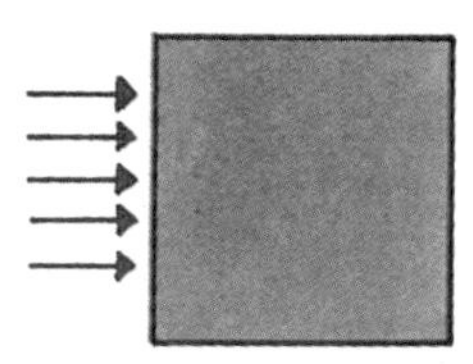

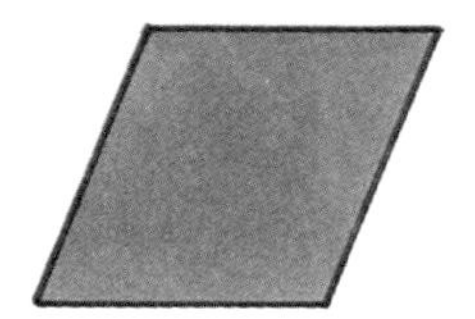

simple shear strain

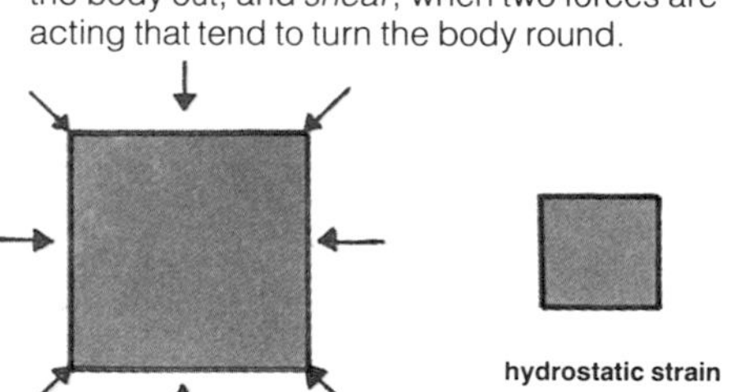

hydrostatic strain

**strain** (*n*) the changes in size and shape produced in rocks and other materials by stress (↑).
**deformation** (*n*) any change in the shape of a mass of rock, whether large or small. **deform** (*v*).
**fracture** (*n*) a break in a rock caused by deformation (↑). **fracture** (*v*), **fractured** (*adj*).

**dip** (*n*) the angle that a bedding-plane (p.80) or other surface on or in a rock makes with the horizontal. *True dip* is measured at 90° to the strike (↓); *apparent dip* is the angle as measured in any other direction (e.g. in a vertical section).

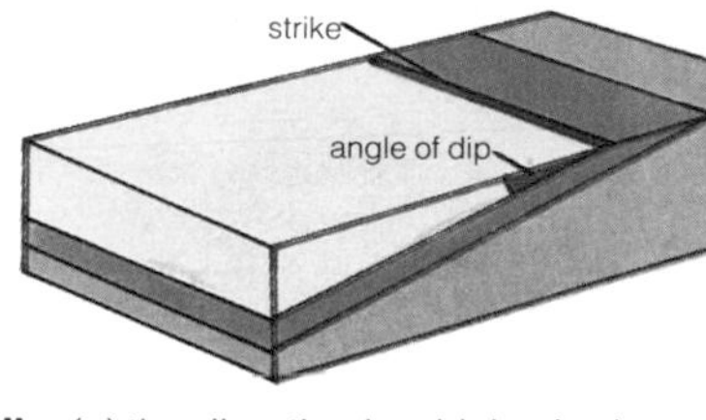


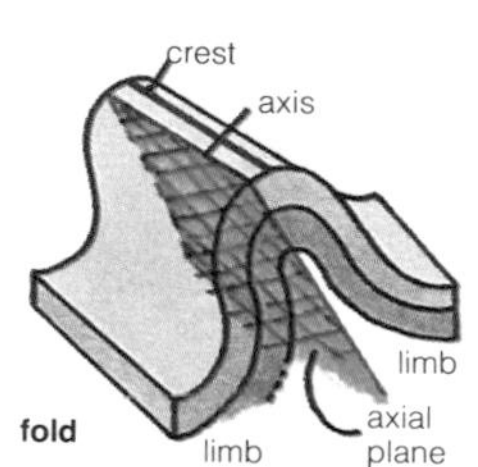


fold

**strike** (*n*) the direction in which a horizontal line can be drawn on a bedding-plane (p.80) or other structural surface at any particular point. The strike is at 90° to the dip (↑).

**fold** (*n*) a bend in a rock mass such as a bed in a series of strata rocks (p.80). **fold** (*v*), **folded** (*adj*).

**axial plane** an imaginary plane that divides a fold (↑) into two more or less equal halves.

**fold-axis** (*n*) an imaginary line that passes through the points where the axial plane (↑) of a fold (↑) cuts a bedding surface (p.80).

**limb** (*n*) one of the two sides of a fold (↑).

**hinge** (*n*) the part of a fold (↑) where it is most sharply curved.

**plunge** (*n*) if the axis (↑) of a fold (↑) is not horizontal it is said to *plunge*. The amount of plunge is the angle between the fold-axis (↑) and the horizontal, as measured in the vertical plane. **plunging** (*adj*).

**pitch** (*n*) the angle between a line (e.g. a fold-axis) in a plane and the horizontal as measured in the plane containing the line (e.g. the axial plane of the fold); (2) = plunge (↑). **pitching** (*adj*).

**closure** (*n*) the direction of closure of a fold (↑) is the direction in which the limbs (↑) become closer together. **close** (*v*).

**quaquaversal** (*adj*) pointing away from a central point in all directions. A word used especially to describe dip (↑); e.g. a dome (p.125) has quaquaversal dip.

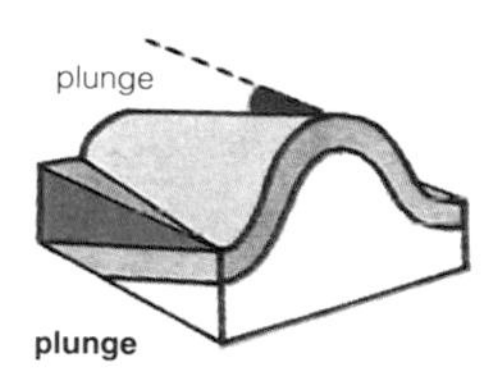


plunge

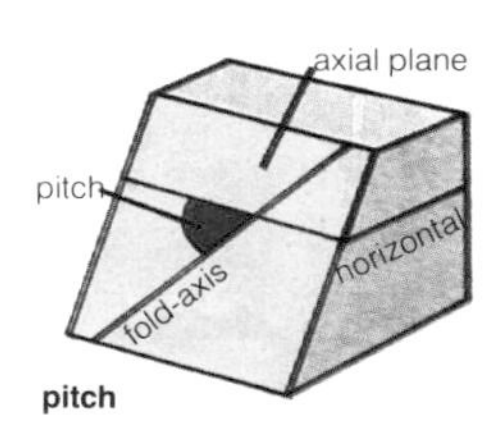


pitch

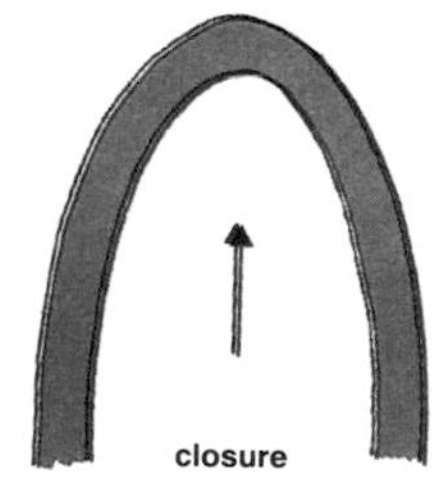

closure

**anticline** (*n*) a fold (p.123) shaped like an arch; a fold in which older rocks are in the centre and younger rocks are outside them. *See also* **syncline** (↓). **anticlinal** (*adj*).
**syncline** (*n*) a fold (p.123) shaped like the letter U; the opposite of an anticline (↑). **synclinal** (*adj*).

**monocline** (*n*) a sharp bend in a bed (p.113) that has the same dip on either side of the bend. **monoclinal** (*adj*).
**anticlinorium** (*n*) a large anticline (↑) which is made up of smaller folds (p.123).
**synclinorium** (*n*) a large syncline (↑) which is made up of smaller folds (p.123).
**fan fold** an anticlinal fold (↑) in which the limbs dip (p.123) towards each other or a synclinal fold (↑) in which the limbs dip away from each other.
**antiform** (*n*) a fold (p.123) of anticlinal shape (↑). The word 'antiform' can be used for a syncline (↑) that is upside down or where it is not certain whether the beds (p.113) are the right way up. *See also* **synform** (↓). **antiformal** (*adj*).
**synform** (*n*) a fold of synclinal (↑) shape. The word 'synform' can be used for an anticline that is upside down or where it is not certain whether the beds are the right way up. *See also* **antiform** (↑), **synformal** (*adj*).
**crest** (*n*) the highest parts of an antiformal (↑) fold (p.123).
**trough** (*n*) = syncline (↑). *See also* p.133.
**concentric fold** a fold (p.123) in which the bedding-planes (p.80) are a series of circles with the same centre. The thickness of the beds (p.80) (as measured at 90° to the bedding-planes) is the same in all parts of the fold.
**parallel fold** = concentric fold (↑).
**similar fold** a fold (p.123) in which the bedding-planes (p.80) are of the same shape. The beds (p.80) are thus thicker near the hinge (p.123) of the fold and thinner in its limbs (p.123).

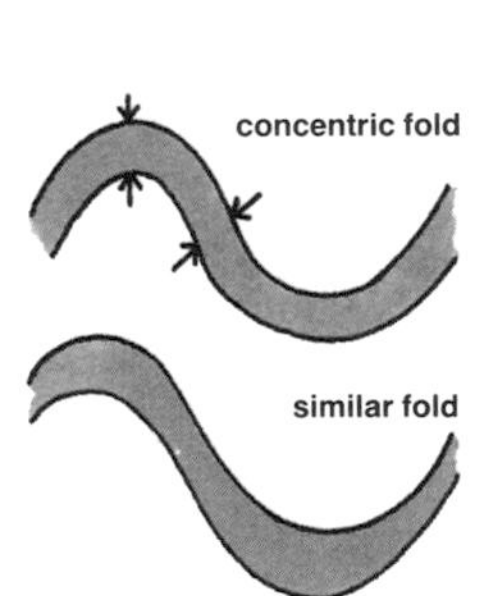

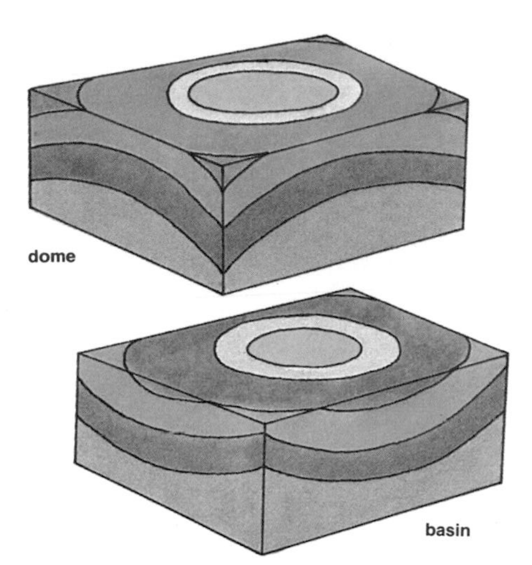


**basin, structural** a structure of generally round shape in which the beds dip (p.123) inwards towards the centre, with younger beds in the centre.

**dome** (*n*) an anticlinal (↑) structure in which the beds dip (p.123) outwards in all directions.

**pericline** (*n*) a general term for domes (↑) and basins (↑). It is more generally used for domes.

**uplift** (*n*) upward movement of a large area of the Earth's crust (p.9). *See also* **subsidence** (↓). **uplift** (*v*).

**subsidence** (*n*) a sinking of a large area of the Earth's crust (p.9); the opposite of uplift (↑). **subside** (*v*).

**warping** (*n*) gentle bending of the Earth's crust (p.9). **warp** (*v*), **warp** (*n*).

**upwarp** (*n*) a large area of the Earth's crust (p.9) that has been uplifted (↑), usually as a broad anticline (↑). *See also* **downwarp** (↓).

**downwarp** (*n*) a large area of the Earth's crust (p.9) that has moved downward, usually as a broad syncline (↑); the opposite of upwarp (↑).

**epeirogenic movements** upward or downward movements of large areas of the Earth's crust (p.9) without folding (p.123).

zig-zag fold

**zig-zag fold** a fold (p.123) in which the limbs (p.123) are straight and the hinges (p.123) are sharp bends; its general shape is thus like a letter Z.

**chevron fold** = zig-zag fold (↑).

**concertina fold** = zig-zag fold (↑).

**disharmonic fold** a fold (p.123) in which the beds do not show a regular arrangement.

**symmetrical fold** a fold (p.123) in which the two limbs (p.123) dip (p.123) at about the same angle. *See also* **asymmetrical fold** (↓).

**asymmetrical fold** a fold (p.123) in which the two limbs do not dip at the same angle. *See also* **symmetrical fold** (↑).

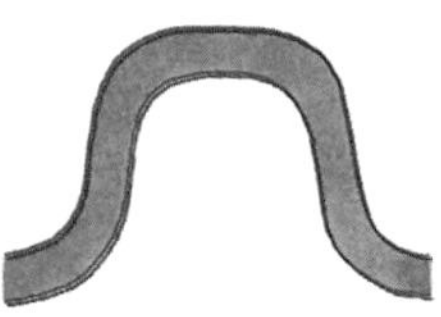

box fold

**cylindrical fold** a fold (p.123) in which the bedding-planes (p.80) are shaped like a drum or a tube.

**box fold** a fold (p.123) that in cross-section has a square shape.

**shear fold** a fold (p.123) formed by small movements along closely spaced cleavage planes (p.95) or fractures (p.122).

**slip fold** = shear fold (↑).

**flow fold** a fold (p.123) formed in incompetent (↓) beds (p.80) that flow like a thick liquid.

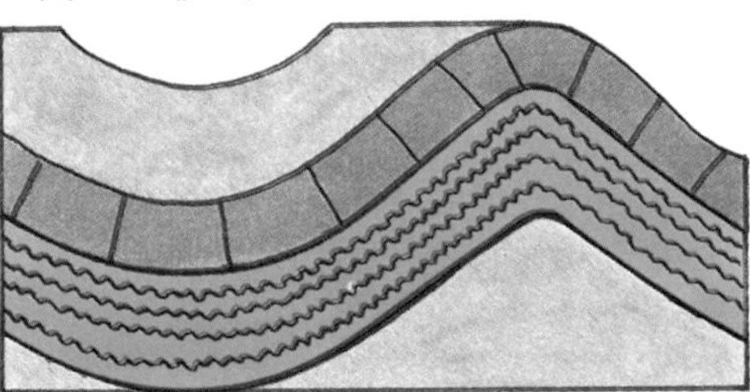

competent

incompetent

**competence**

**competent bed** a bed (p.80) whose thickness remains the same in all places when it is folded. *See also* **incompetent bed** (↓).

**incompetent bed** a bed (p.80) whose thickness varies from place to place when it is folded. *See also* **competent bed** (↑).

**drag fold** a small fold (p.123) formed in an incompetent bed (↑) by slip (p.128) along the bedding-planes (p.80).

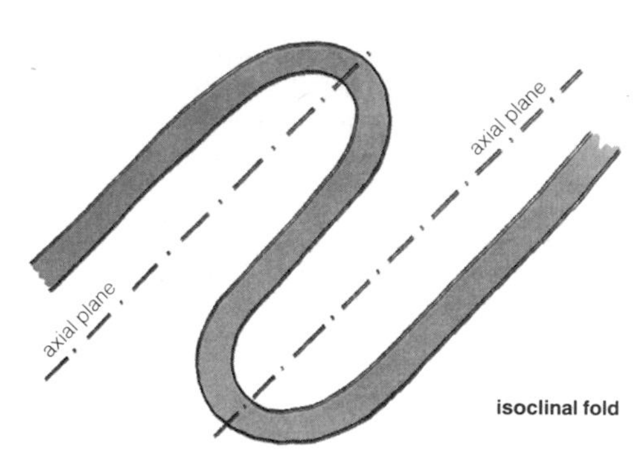

isoclinal fold

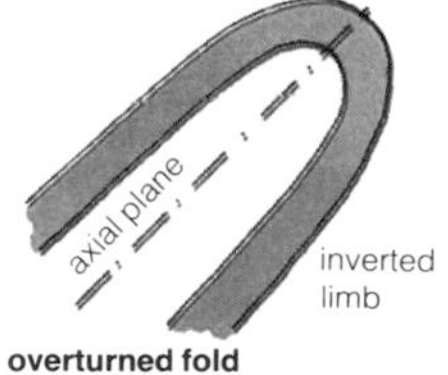

overturned fold

**isoclinal fold** a fold (p.123) in which the two limbs (p.123) are parallel.

**overturned fold** a fold (p.123) in which the axial plane (p.123) is so far from the vertical that one limb (p.123) is over part of the other limb.

**recumbent fold** a fold (p.123) in which the axial plane (p.123) is horizontal or nearly horizontal. *See also* **nappe** (p.130).

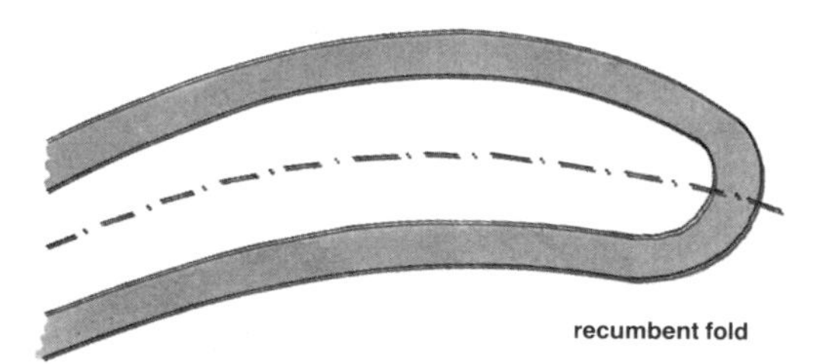
recumbent fold

**inverted** (*adj*) upside down. **inversion** (*n*).

**uninverted** (*adj*) the right way up.

**way up** strata (p.80) are the right way up if the youngest beds (i.e. those deposited (p.80) last) are on top. Strata are inverted (↑) if they have been overturned (↑).

**young** (*v*) beds (p.80) are said to 'young' in the direction in which the youngest surface faces. The verb 'to young' is useful in describing the relationships of beds in areas of complex folding (p.123).

**face** (*v*) = young (↑).

**fault** (*n*) a break in the rocks along which movement has taken place.

**fault plane** the surface along which fault movement (↑) has taken place. A fault-plane may be a smooth surface or a broad zone. *See also* **fault zone** (↓).

**fault zone** a broad area along which fault movement (↑) has taken place. A fault zone may be several hundred metres wide. *See also* **fault plane** (↑).

**upthrow** (*adj*) the side of a fault (↑) on which the throw (↓) is upward in relation to the other side. *See also* **downthrow** (↓).

**downthrow** (*adj*) the side of a fault on which the throw (↓) is downward in relation to the other side. *See also* **upthrow** (↑).

**hanging wall** the rocks that lie above the fault plane (↑) of a fault that is not vertical. *See also* **footwall** (↓).

**footwall** (*n*) the rocks that lie below the fault plane (↑) of a fault that is not vertical. *See also* **hanging wall** (↑).

**throw** (*n*) the amount of movement that has taken place along a fault (↑) as measured in the vertical direction. *See also* **heave** (↓).

**heave** (*n*) the amount of horizontal movement that has taken place between the two sides of a fault (↑). *See also* **throw** (↑).

**hade** (*n*) the angle between a fault plane (↑) and the vertical. It is equal to 90° minus the angle of dip (p.123) of the fault.

**displacement** (*n*) the distance between two points that were next to each other before fault (↑) movement took place.

**offset** (*n*) the horizontal displacement (↑) of a fault (↑), measured parallel to the strike (p.123) of the fault.

**net slip** = displacement (↑).

**slip** = net slip (↑).

**normal fault** a fault (↑) in which the hanging wall (↑) has moved downward in relation to the footwall (↑). *See also* **reverse fault** (↓).

**reverse fault** a fault (↑) in which the hanging wall (↑) has moved upward in relation to the footwall (↑). *See also* **normal fault** (↑).

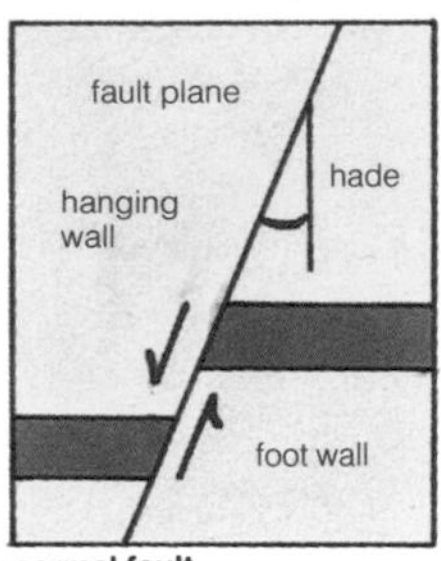

**normal fault**

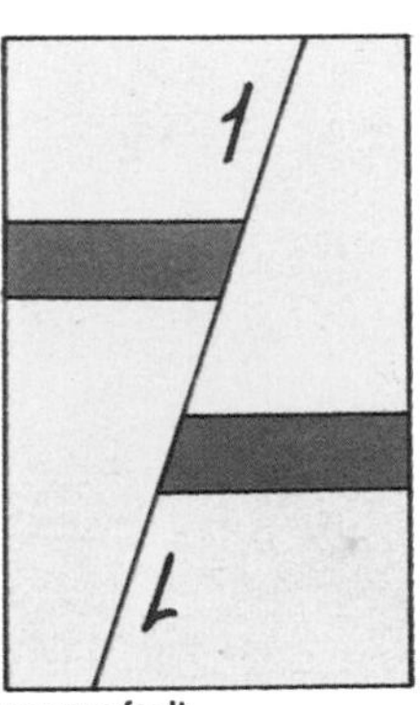
**reverse fault**

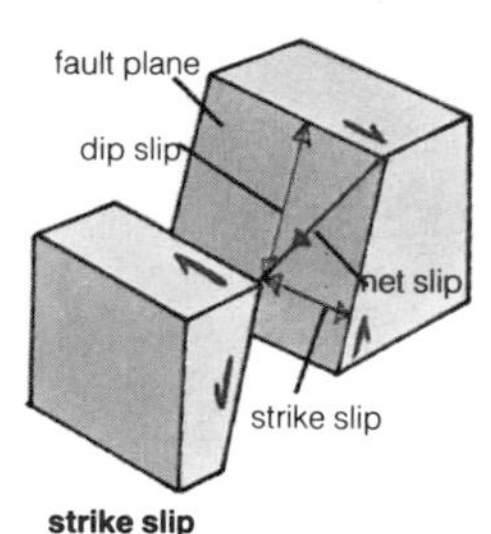

**strike slip**

**strike-slip fault** a fault (↑) in which the movement on the fault plane (↑) is parallel to the strike (p.123) of the fault, i.e. sideways.

**wrench fault** a steeply dipping (p.123) fault (↑) on which the movement has been horizontal, i.e. strike-slip (↑).

**tear fault** (1) a wrench fault (↑); (2) a strike-slip fault (↑) that crosses the strike (p.123).

**transcurrent fault** = tear fault (↑).

**dextral fault** a strike-slip fault (↑) in which the relative displacement (↑) is to the right as seen across the fault plane (↑). *See also* **sinistral fault** (↓).

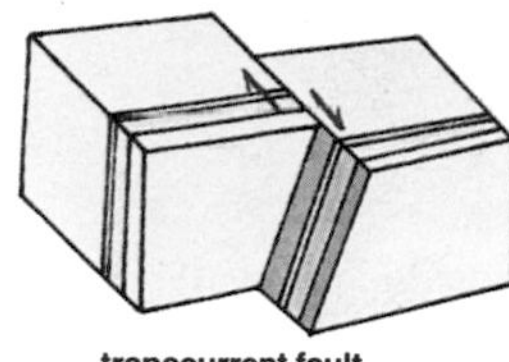
**transcurrent fault (dextral)**

**sinistral fault** a strike-slip fault (↑) in which the relative displacement (↑) is to the left as seen across the fault plane (↑). *See also* **dextral fault** (↑).

**oblique-slip fault** a fault (↑) in which the displacement (↑) is not parallel to the strike (p.123) or to the dip (p.123) of the fault plane.

**rotational fault** a fault (↑) in which one block (p.133) has turned about a point on the fault plane (↑). The displacement (↑) varies from place to place.

**dip-slip fault** a fault (↑) in which the movement on the fault plane (↑) is parallel to the dip (p.123) of the fault.

**slickenside** (*n*) (*usually in the plural, slickensides*) fine parallel scratches or grooves on a fault (↑) surface that have been produced by the movement of the rocks on either side of the fault.

**fault breccia** a breccia (p.87) made up of sharp pieces of rock broken up during movement along a fault (↑).

**fault gouge** finely divided rock material produced by movement along a fault (↑).

**thrust fault** a reverse fault (↑) in which the fault plane (↑) is at a low angle to the horizontal.

**thrust** (*n*) = thrust fault (↑).

**thrust plane** the plane of a thrust fault (↑).

**overthrust** (*n*) = thrust fault (↑).

**fault block** a mass of rock with faults (↑) on two or more sides.

**fault scarp** an escarpment (p.33) produced by movement on a fault (↑).

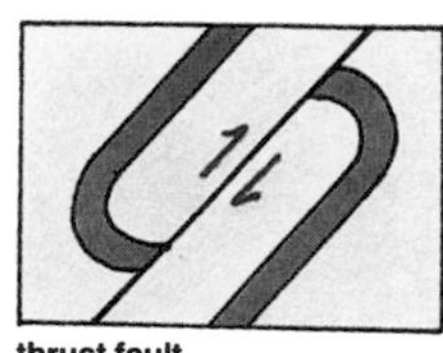
**thrust fault**

**listric fault** a fault (p.128) that curves downwards, the fault plane (p.128) being steep at the surface and more nearly horizontal at depth. Listric faults are characteristic of continental margins (p.135).

**lag** (*n*) a thrust fault (p.129) in which the uninverted (p.127) limb (p.123) of a recumbent fold (p.127) has been cut out.

**slide** (*n*) a fault (p.128) that is nearly horizontal. A slide may be a thrust fault (p.129) or a lag (↑).

**nappe** (*n*) a large mass of rock that has been moved several kilometres or more. A nappe may be either the hanging wall (p.128) of an overthrust (p.129) or a recumbent fold (p.127).

**decke** (*n*) **(decken)** = nappe (↑).

**thrust sheet** = nappe (↑).

**imbricate structure** a structure in which reverse faults (p.128) are formed between thrust planes (p.129) that are more or less parallel.

**schuppen structure** = imbricate structure (↑).

**décollement** (*n*) 'unsticking'. A series of beds (p.80) may be folded (p.123) and slide (↑) over a lower series that is little folded or not folded at all; this is a décollement. It is necessary for there to be a bed at the base of the folded series that can slip easily over the rocks below, e.g. salt (p.17) or anhydrite (p.52) beds.

**nappe**

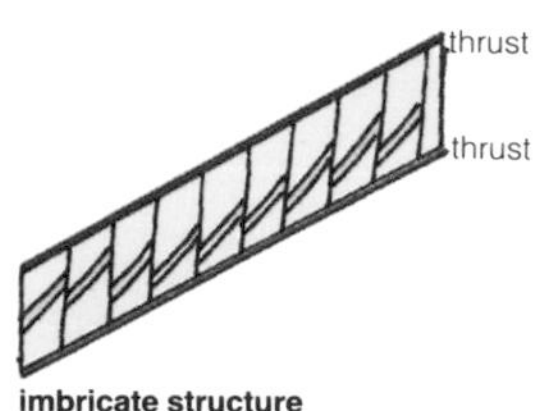


**imbricate structure**

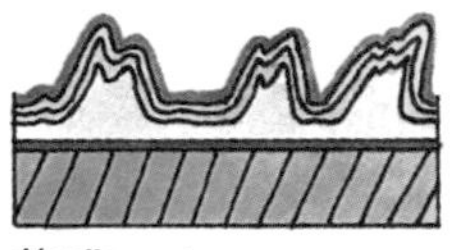

**décollement**

**klippe**

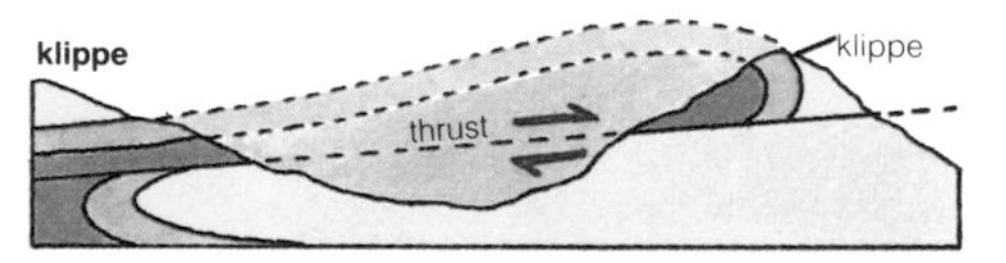


**klippe** (*n*) (*klippen*) a piece of a nappe (↑) that stands apart and is separated from the other rocks of the nappe and rests on a thrust plane (p.129), thus forming a *tectonic outlier* (p.118).

**window** (*n*) an area in which the rocks above a thrust fault (p.129) have been eroded to expose (p.122) the rocks below.

**fenster** (*n*) = window.

**gravity tectonics** the movement of rocks under the force of gravity (p.11) to produce tectonic structures (p.122) such as faults (p.128) and folds (p.123).

**diapir** (*n*) a structure (p.122) in which the core (p.9) of an anticline (p.124) breaks through the rocks above. Salt (p.17) can form diapirs; so can igneous (p.62) rocks.

**salt dome** a diapir (↑) of salt (sodium chloride).

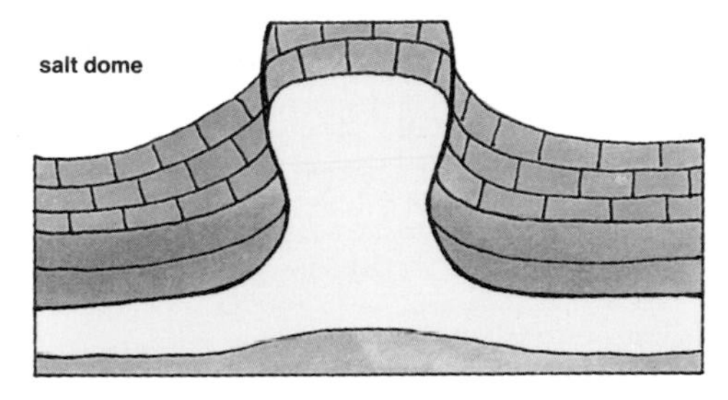


**piercement dome** a salt dome (↑) in which the mass of salt has broken through the rocks above it and has reached or nearly reached the Earth's surface.

**mantled gneiss dome** a dome (p.125) of granite (p.76) surrounded by gneiss (p.97) and sediments (p.80). The foliation (p.95) of the granite is parallel to the bedding (p.80) of the sediments where they are in contact.

**cryptoexplosion structure** a term used to describe more or less circular structures that show intense rock deformation (p.122) but do not appear to have been caused by volcanic or tectonic activity. These structures range from 1.5 km to more than 50 km in diameter. They have evidently been formed by an explosive force but the term 'cryptoexplosion structure' does not imply any particular method of formation. *See also* **astrobleme** (↓).

**astrobleme** (*n*) an ancient mark on the surface of the Earth produced by the fall of a large body from space. Astroblemes are usually circular and the rocks show signs of great shock. *See also* **cryptoexplosion structure** (↑).

**lineament** (*n*) any geological structure of large size that appears as a geographical feature in the form of a straight line, e.g. a valley, a hill, or a coastline.

**diastrophism** (*n*) the deformation (p.122) of large masses of the Earth's crust (p.9) to form mountains, etc. **diastrophic** (*adj*).

**orogeny, orogenesis** (*n*) the process of mountain-building; a period of mountain-building. Deformation (p.122), folding (p.123), and thrusting (p.129) are characteristic features of orogenies and are usually accompanied by the intrusion (p.64) of igneous rocks (p.62). The period of time required for an orogeny may be hundreds of millions of years. **orogenic** (*adj*).

**orogen, orogenic belt** a relatively narrow region, which may be thousands of kilometres long and hundreds of kilometres wide, that has been affected by an orogeny (↑). The deeper parts are affected by regional metamorphism (p.90) and the emplacement (p.64) of igneous rocks (p.62).

**mobile belt** a long, narrow region of the Earth's crust (p.9) in which there is deformation (p.122), igneous activity (p.62), and metamorphism (p.90). A mobile belt will have stable blocks (↓) on either side of it.

**geosyncline** (*n*) a long, narrow area of the Earth's crust (p.9) in which a great thickness of sediment (p.80) is deposited. Volcanic rocks (p.68) are also present with the sediments. The floor of the geosyncline subsides (p.125) as sedimentation continues. The sediments may later be deformed (p.122) to produce a mountain range. **geosynclinal** (*adj*).

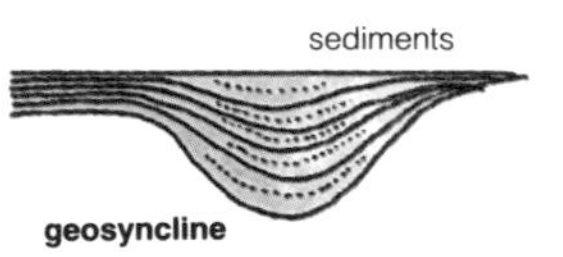


**miogeosyncline** (*n*) a geosyncline (↑) formed next to a craton (↓). The sediments (p.80) in a miogeosyncline are relatively thin and there are no volcanic rocks (p.68). **miogeosynclinal** (*adj*).

**eugeosyncline** (*n*) a geosyncline (↑) formed away from a craton (↓) and containing a great thickness of sediments (p.80), including greywackes (p.87) and volcanic rocks (p.68). **eugeosynclinal** (*adj*).

**geanticline** (*n*) an uplift of anticlinal (p.124) form developed in a geosyncline (↑) as the sides of the geosyncline move closer together. **geanticlinal** (*adj*).

**cordillera** (*n*) (1) a mountain range or a series of more or less parallel mountain ranges; (2) a row of islands in a geosyncline (↑), formed when the rocks of a geanticline (↑) reach sea level.

**block** (*n*) in tectonics (p.122), a large mass of the Earth's crust, tens or hundreds of kilometres across, that behaves as a single rigid unit.

**flexure** (*n*) a bend in a strata (p.80), usually a gentle one; a fold (p.123). **flexural** (*adj*).

**taphrogenesis** (*n*) vertical movements of large size that produce steeply dipping faults (p.128).

**shield** (*n*) a large area of very old igneous (p.62) and metamorphic (p.90) rocks of Pre-Cambrian (p.114) age that have not been folded or deformed (p.122) since Pre-Cambrian times.

**platform** (*n*) a large area of old, stable basement rocks (p.114) covered by younger horizontal or nearly horizontal strata (p.80) resting on the eroded (p.20) surface of the basement.

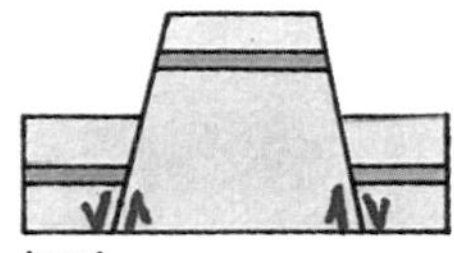

horst

**foreland** (*n*) a resistant block (↑), i.e. a block that is stable and rigid, on one side of a geosyncline (↑); the side of a folded mountain range towards which the overturned folds (p.123) are leaning.

**craton** (*n*) a stable area of the Earth's crust (p.9), usually large. More or less the same as a shield (↑).

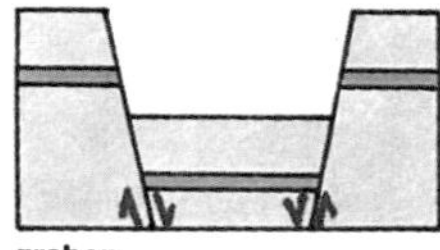

graben

**horst** (*n*) an area that has been lifted up as a block between normal faults (p.128) on either side of it.

**graben** (*n*) (*graben*) a narrow block of the Earth's crust (p.9) that has been moved down between normal faults (p.128) on either side. *See also* **rift valley** (↓).

**rift** (*n*) a structural feature in which the rocks between two faults (p.128) have been let down in relation to those on either side. The use of the term 'rift' does not necessarily mean that there is a valley in the geographical sense. *See also* **rift valley** (↓).

**trough faulting** = rift (↑).

**rift valley** a valley formed between two more or less parallel faults (p.128).

**aulacogen** (*n*) a trough (p.133) formed by a rift (↑) that has failed to develop.

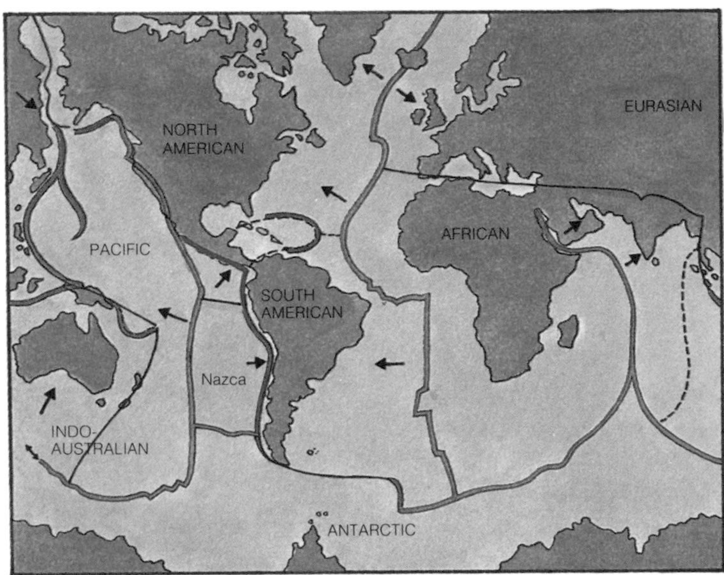


**the major plates**

spreading ridge

subduction zone

movement of plate

**plate tectonics** the theory that the Earth's surface is covered by a number of relatively thin plates (↓), which move over the material below. Many geological facts that earlier appeared to be unrelated have been explained by this theory.

**plate** (*n*) part of the Earth's surface that behaves as a single rigid unit. Plates are about 100 to 150 km thick. They may be made up of continental crust or oceanic crust (p.9), or both, on top of a layer of the upper mantle (p.9). Plates move in relation to the Earth's axis and to each other. There are seven large plates (the African, Eurasian, Indo-Australian, Pacific, North American, South American, and Antarctic plates) and several smaller ones.

**microplate** (*n*) a small plate (↑), e.g. those round the Mediterranean Sea.

**continental drift** the theory that the present continents have been formed by the breaking up of one large continent and have since moved to their present positions.

**plate margin** the edge of a plate (↑). It is at the plate margins that most seismic (p.12), volcanic (p.68), and tectonic (p.122) activity is found. There are three types: constructive margins, at which new crust (p.9) is being formed; destructive margins, at which one plate is moving down below another; and conservative margins, at which plates simply move past each other.

**constructive margin** see plate margin (↑).

**destructive margin** see plate margin (↑).

**conservative margin** see plate margin (↑).

**plate boundary** the line between two plates (↑) that touch each other. Plate boundaries are marked by seismic activity (p.12) and tectonic activity (p.122).

**triple junction** a point on the Earth's surface where three plate boundaries (↑) meet.

**island arc** a curved chain of islands with the convex (outer) side of the curve facing the open ocean. There is a deep oceanic trench (p.35) on the convex side of the arc and deep sea on the opposite side. Island arcs are regions where deep-focus earthquakes (p.12) occur and where gravity (p.11) and magnetic anomalies (p.14) are found. The islands may also show volcanic activity (p.68).

**island arc:**
the Kuril island arc, north of Japan

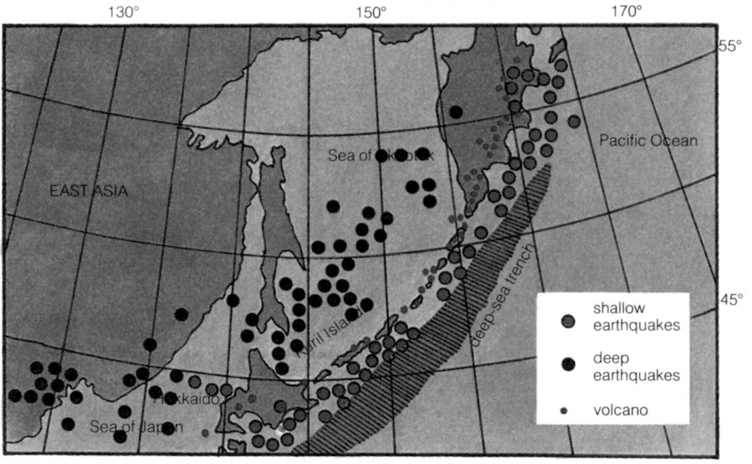

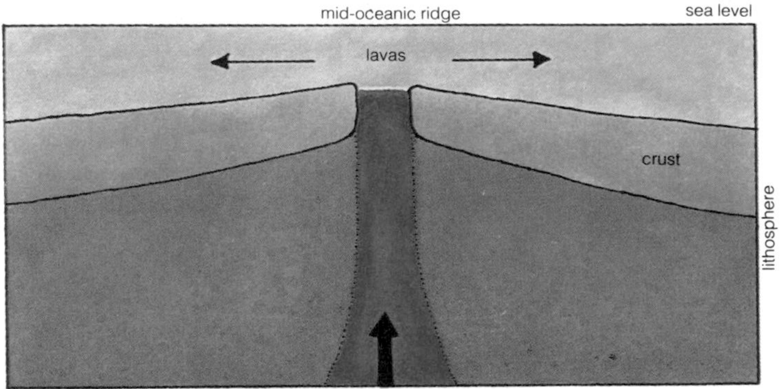

**sea-floor spreading**

**sea-floor spreading** the theory that growth of new crust (p.9) takes place at active mid-oceanic ridges (p.35). This takes place by the intrusion (p.64) of submarine lavas (p.70) at the mid-oceanic ridges. The rocks nearest to the ridge are thus the youngest and the age of the rocks on the sea floor increases with distance from the axis of the ridge. Magnetic anomalies (p.14) shown by the rocks on the two sides of the ridge are symmetrical about the ridge axis.

**transform fault** a fault (p.128) along which two plates (p.134) move past each other without lithosphere (p.9) being formed or destroyed. A typical transform fault is a strike-slip fault (p.129) that cuts across a mid-oceanic ridge (p.35), the ridge being offset (i.e. its two halves do not meet across the fault). There is seismic activity (p.12) at the transform fault between the two points where it meets the mid-oceanic ridge. There are also continental transform faults, e.g. the San Andreas fault in California and the North Anatolian transform fault south of the Black Sea.

**zone of divergence** a constructive margin (p.135); a region where two plates (p.134) are moving away from each other, e.g. the Mid-Atlantic Ridge. New lithospheric (p.9) material is formed in these regions.

**pull-apart zone** = zone of divergence (↑).

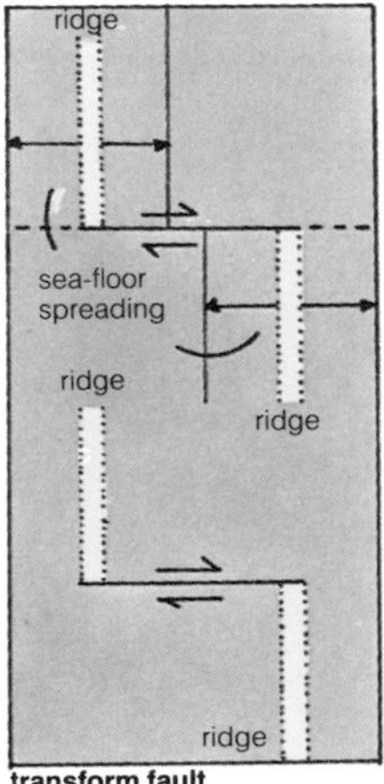

**transform fault**

**subduction zone** a region in which a lithospheric plate (pp.9, 134) is forced down, or subducted, into the asthenosphere (p.9) and mesosphere (p.9). The movement of the lithospheric plate is thought to be the cause of the earthquakes (p.12) that occur in island arc regions (p.135). As it moves down into the mantle (p.9) the plate is heated, and at a depth of between 100 and 300 km it is partly melted. At 700 km depth it breaks up completely. *See also* **Benioff zone** (↓).

**Benioff zone** a sloping surface of seismic activity (p.12) that is characteristic of island-arc (p.135) systems. The Benioff zone meets the Earth's surface close to an ocean trench (p.35) and dips below the island arc. The angle of dip is typically about 45° but may be between 30° and 80°.

**zone of convergence** = subduction zone (↑).

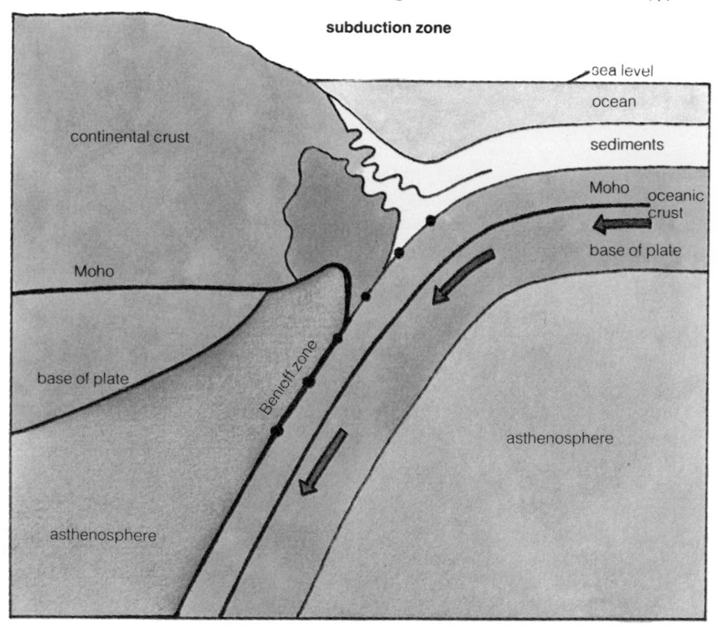

**accretionary prism** a mass of sediments (p.80) pushed together at a subduction zone (p.137). Older sediments are thrust (p.129) over younger sediments as one plate is driven under the other. (*See diagram on p.137.*)

**back-arc upwelling** island arc systems (p.135) tend to move away from the continent and into the ocean. This is thought to be caused by convection currents (p.142) in the asthenosphere (p.9) between the island arc and the continent, which in turn are caused by cooling of the mantle (p.9) by the descending mass of oceanic lithosphere (p.9).

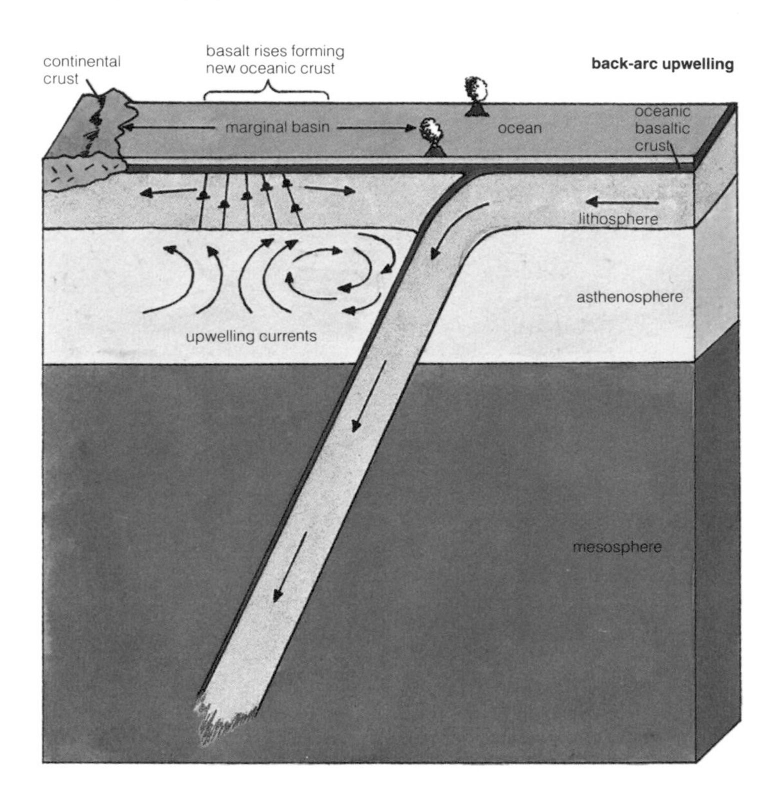

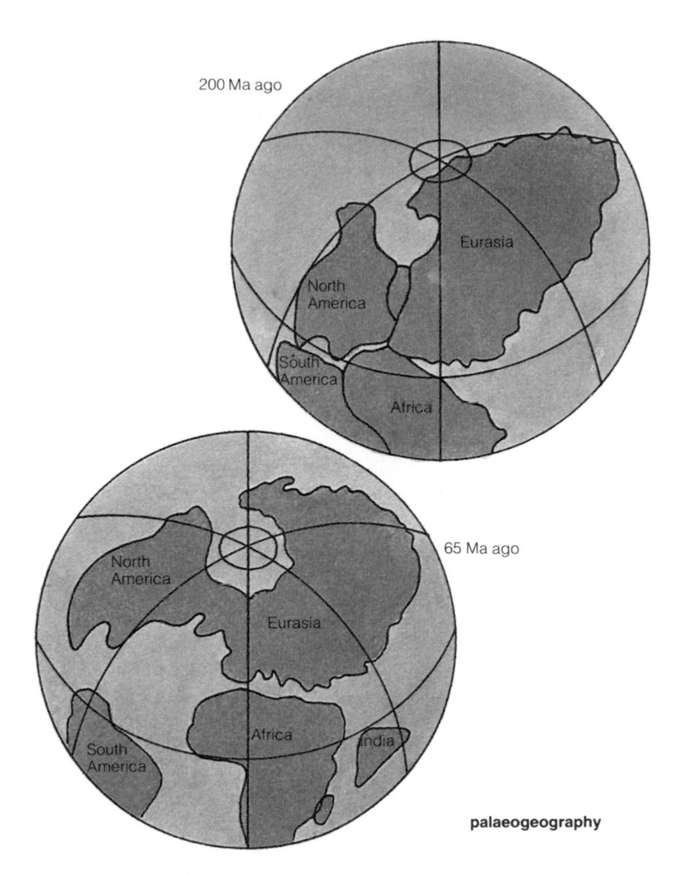


**palaeogeography**

**palaeogeography** (*n*) the study of the geography of past geological ages and especially of the positions of former continents and oceans. This is done by studying rocks of known ages to discover the environments (p.81) in which they were formed and the directions in which the rivers and the ocean currents were flowing at the time. The information is then put together, usually in the form of a *palaeogeographical map*. **palaeogeographical, palaeogeographic** (*adj*).

**Gondwanaland** (*n*) the 'supercontinent' that is thought to have existed in the southern hemisphere until the Cretaceous (p.115). It consisted of South America, Africa, Arabia, Madagascar, India, Sri Lanka, Australia, New Zealand, and Antarctica. *See also* **Pangaea** (↓).

**Laurasia** (*n*) the 'supercontinent' that is thought to have existed in the northern hemisphere at some time before the Tertiary (p.115). It consisted of North America, Greenland, and Eurasia (Europe and Asia). *See also* **Pangaea** (↓).

**Pangaea** (*n*) the 'supercontinent' formed by Gondwanaland (↑) and Laurasia (↑) together. Pangaea began to break up about 200 Ma ago in the Jurassic (p.115).

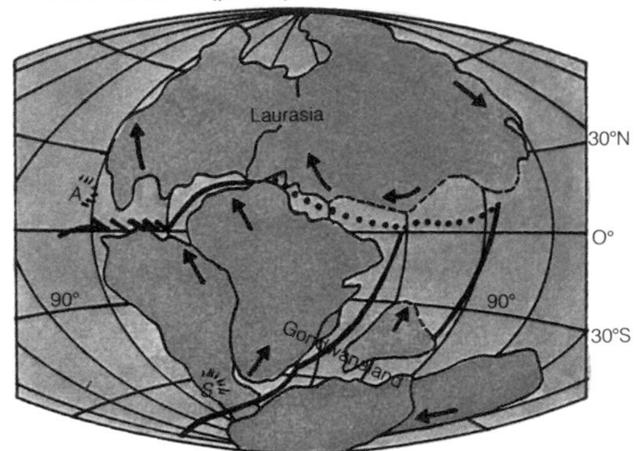


**Laurasia and Gondwanaland 180 Ma ago**

**continental accretion** the growth of continents by adding material to their margins. The central parts of the continents are very old: Pre-Cambrian (p.114). Younger rocks have, it is thought, been added to these central parts by accretion during mountain-building periods when the original continents have come together as a result of plate movements (p.134).

**suture** (*n*) a belt of deformed rock (an orogenic belt, p.132), which marks the zone where two continents have come together and joined. The suture may be several hundred kilometres wide.

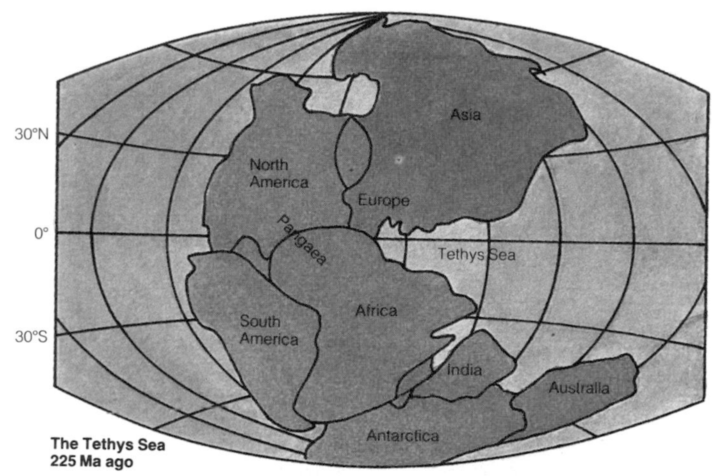


**The Tethys Sea 225 Ma ago**

**Tethys** (*n*) the ocean that is thought to have existed between the eastern ends of Gondwanaland (p.140) and Laurasia (p.140). The Tethys sea was formed when Gondwanaland moved south between 340 and 225 Ma ago. The eastern end of the Tethys sea was closed in the Tertiary (p.115), leaving the present Mediterranean Sea between Europe and Africa.

**Iapetus sea** a sea that is thought to have separated the English and European continent from Scotland and North America in Lower Palaeozoic times.

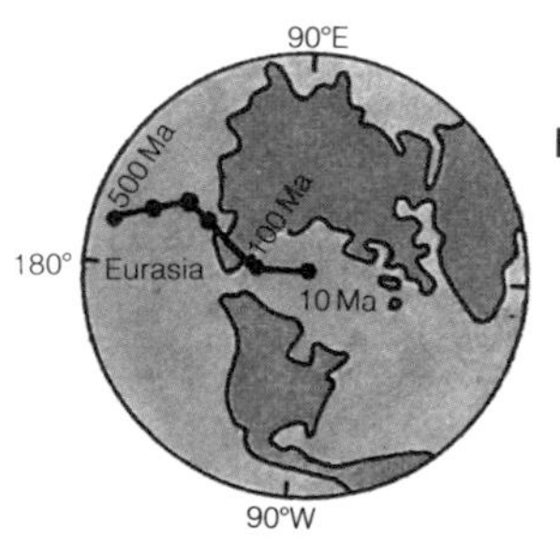


**apparent polar wandering path for Eurasia**

**polar wander** when the palaeomagnetism (p.14) of rocks of various ages is studied and their directions of magnetization (p.14) are measured, the positions of the Earth's magnetic poles (p.14), as shown by the magnetism of the older rocks, are not close to the Earth's geographical poles. Lines called *polar wandering curves* can be drawn on maps to show the apparent movements of the poles in the geological past. The apparent movement of the poles can be explained by movements of the continents, and this explanation fits in with the theory of continental drift (p.134).

**convection current** convection could occur in the Earth's mantle (p.9) if the material in the upper part of the mantle were cooled. This material would then descend and its place would be taken by hotter material from below, thus setting up a convection current. Currents of this kind could provide one possible means of moving the lithospheric plates (pp.9, 134).

**plume** (*n*) it is thought that hotter material could move upward in the Earth's mantle (p.9) in a number of *thermal plumes*, each a few hundred kilometres across. On reaching the top of the mantle, the hotter material would spread out sideways in all directions and would be cooled. There would then be a downward flow of material in other regions to balance the upward flow in the plumes. The sideways movement of material could provide a means of moving the lithospheric plates (pp.9, 134), and the plumes could explain the occurrence of chains of volcanic islands (p.68) that have no relationship to plate boundaries (p.134). *See also* **hot spot** (↓).

**hot spot** an area of the lithosphere (p.9) that is heated by a plume (↑). Volcanic activity (p.68) and upward doming (p.125) of the lithosphere could be caused by a hot spot. A line of volcanoes (p.68) could be formed when a plate (p.134) moved over a fixed hot spot.

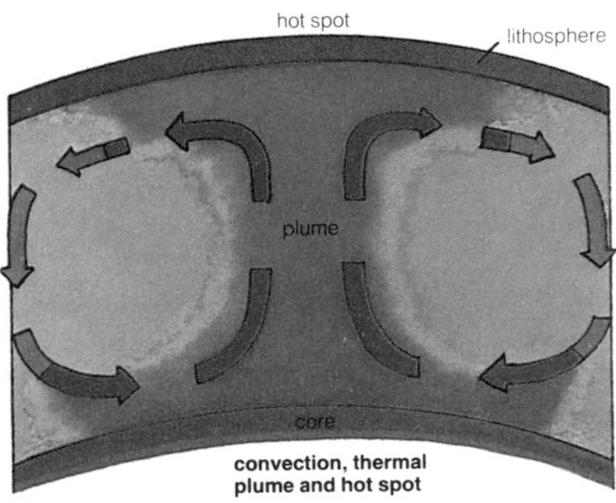

**convection, thermal plume and hot spot**

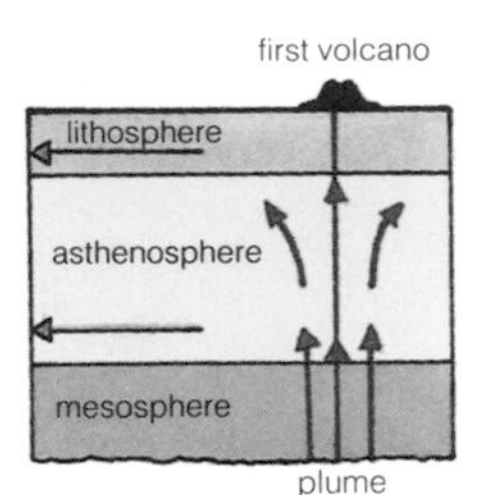


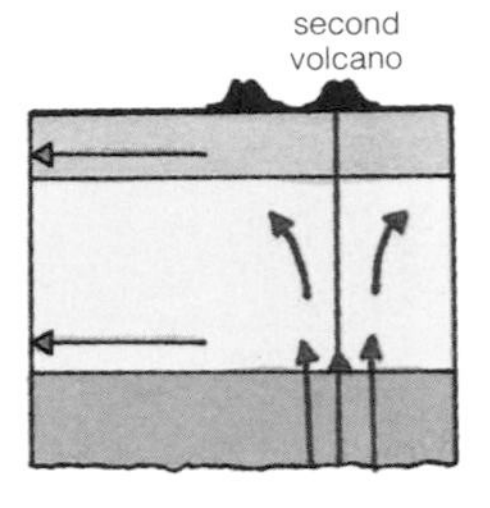


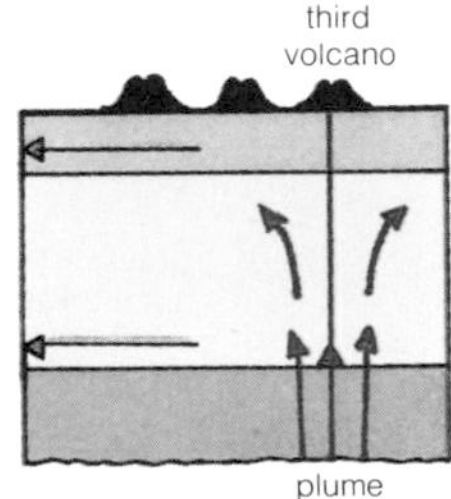

**formation of a line of volcanoes by a fixed hot spot**

**rock mechanics, soil mechanics** the study of the physical properties of rocks and soils, especially those properties that affect their ability to support a load.

**geotechnics** = soil mechanics (↑).

**site investigation** the geological study of a piece of ground on which a building or some other structure (e.g. a dam) is to be placed.

**soil creep** the very slow movement of soil (p.23) at the surface down a slope.

**earth flow** movement of surface material that is faster than soil creep (↑) but slower than a mud flow (↓) or landslide (↓). An earth flow usually slides on a spoon-shaped surface. At its upper end is a curved cliff and at its lower end is a swelling shaped like a tongue.

**mud flow** the rapid movement of a mixture of mud and water, which flows like a liquid. A typical mud flow can carry large rocks and boulders. Mud flows are characteristic of desert regions and alpine regions (i.e., the higher regions of mountain systems).

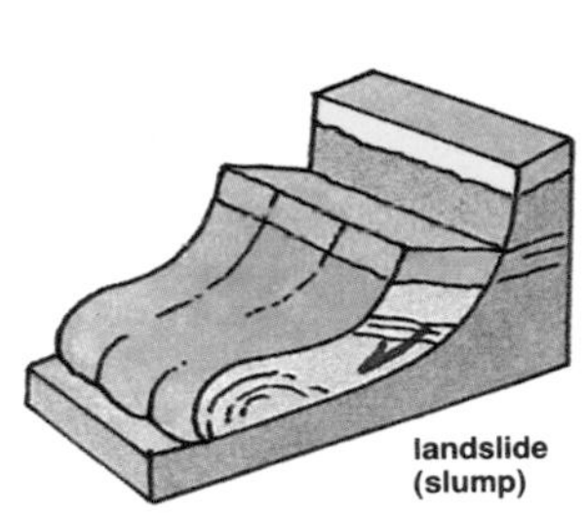

**landslide (slump)**

**landslide** (*n*) the movement of a mass of rock or soil, or both, down a slope. A landslide differs from an earth flow (↑) in that the mass of rock remains more or less in one piece.

**landslip** (*n*) = landslide (↑).

**slump** (*n*) a landslide (↑) in which a mass of rock – typically clay – moves on a curved surface (a *shear surface*).

**glide** (*n*) a landslide (↑) in which a mass of rock moves downward along the surface of a sloping bedding plane (p.80).

**rock fall** the free fall of pieces of rock from a cliff or steep slope. Freezing and thawing are a common cause of rock falls. The fallen rocks may pile up as talus cones (p.21) at the foot of the cliff.

**quick clay** a special type of clay (p.88) containing a large amount of water (often more than 50 per cent by weight). Such clays are normally solid but a shock can cause them to turn liquid. They can thus form sudden earth flows. Quick clays were originally deposited on the sea floor close to glaciers (p.28).

**petroleum** (*n*) oil (↓) occurring naturally in the Earth's crust; natural gas (↓), oil (↓), and solid bitumens (p.89).

**oil** (*n*) the oil that occurs naturally in the Earth's crust is a mixture of compounds of carbon and hydrogen (hydrocarbons). It is generally thought to have formed from the remains of plants and animals, or both, but we have no direct knowledge of the way in which oil is formed in the Earth's crust.

**crude oil, crude** oil (↑) as it occurs naturally in the Earth's crust.

**natural gas** gas occurring naturally in the Earth's crust that consists of hydrocarbons (compounds of carbon and hydrogen). Natural gas may be found alone or with oil (↑).

**oil shale** an argillaceous rock (p.85) containing a solid material that when distilled gives off oil. The oil cannot be obtained without distillation.

**migration** (*n*) the upward movement of oil (↑) from the rocks in which it was originally formed to other rocks that are porous (p.84) and permeable (p.145). Pressure (p.93) from the weight of the beds on top of the beds containing the oil can cause it to start to migrate. **migrate** (*v*), **migrating** (*adj*).

**trap** (*n*) something that stops the upward migration (↑) of oil and causes it to accumulate. Traps are of two types: structural (p.122) and stratigraphical (p.112). Examples of structural traps are anticlines, synclines, domes, and faults (pp.116, 117, 120). Stratigraphical traps include unconformities (p.118) and variations in lithology (p.85).

**cap rock** the impermeable (p.146) rock(s) that make a trap (↑) effective in preventing oil from migrating (↑) further.

**reservoir** (*n*) beds in which oil accumulates.

**borehole** (*n*) a hole drilled into the Earth for oil (↑), gas, water, etc. or to gain information about the rocks below the surface.

**well-logging** (*n*) the use of physical measurements from instruments lowered down boreholes (↑) to obtain information about the rocks below the surface.

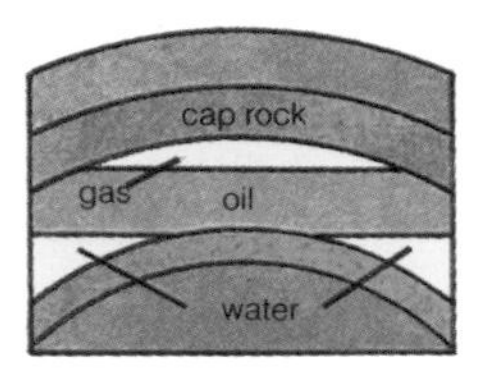


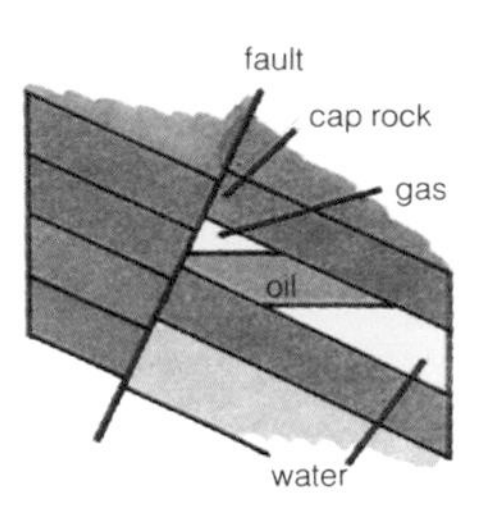


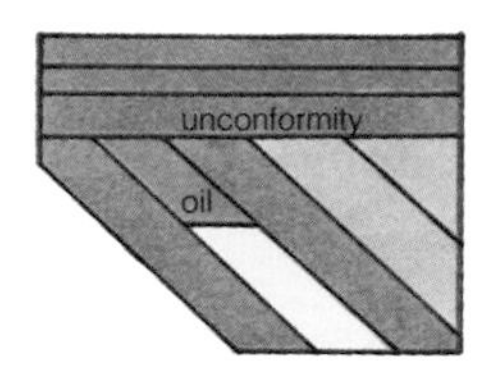


traps

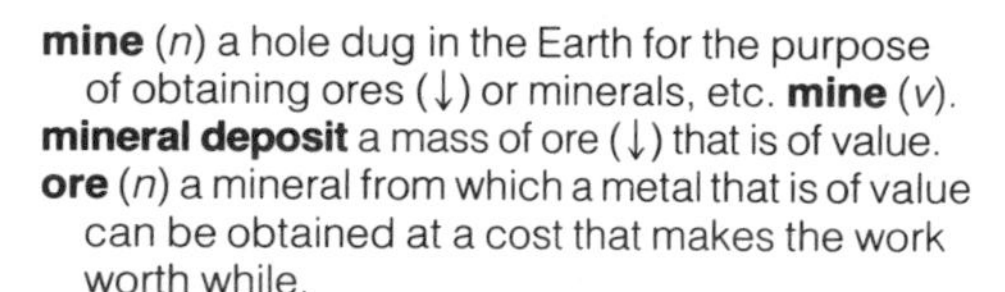

**mine** (*n*) a hole dug in the Earth for the purpose of obtaining ores (↓) or minerals, etc. **mine** (*v*).

**mineral deposit** a mass of ore (↓) that is of value.

**ore** (*n*) a mineral from which a metal that is of value can be obtained at a cost that makes the work worth while.

**ore body** a mass of ore (↑). It may be of hydrothermal (↓) origin or intrusive (p.64): a dyke, sill (p.66), or vein (↓).

**mineralization** (*n*) the formation of new minerals, especially ore minerals, in an existing rock – usually as veins (↓) or masses.

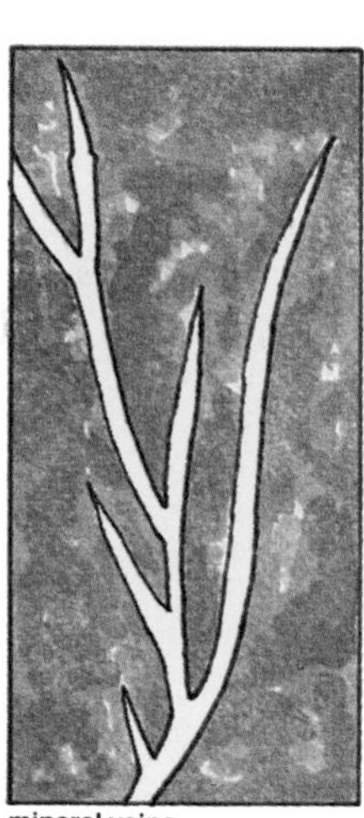

mineral veins

**metallogenetic province, metallogenic province** a region in which there is a series of mineral deposits (↑) having characters in common.

**hydrothermal deposit** a mineral deposit (↑) produced by liquids coming from a magma (p.62) that contain a large proportion of hot water.

**replacement deposit** a mineral deposit (↑) in which the mineral takes the place of a rock that was there earlier, e.g. the hydrothermal (↑) replacement of limestone (p.86) by galena (p.50).

**placer deposit** a deposit, usually at the surface, containing a mineral of value such as gold.

**vein** (*n*) a thin mass of rock or a mineral, especially a thin ore body (↑).

**wall rock** the country-rock (p.65) on either side of a vein (↑).

**lode** (*n*) more or less the same as a vein (↑).

**reef** (*n*) in mining geology, a vein (↑) of quartz (p.55) containing gold.

**gangue** (*n*) the part of an ore (↑) body that does not contain the metal(s) that are being worked.

**stockwork** (*n*) a large mass of rock that is cut across by many small veins (↑).

**opencast** (*adj*) applies to a method of mining (↑) a bed (p.80) or vein (↑) near the surface by cutting into it from above rather than by digging a mine (↑) under the ground.

**overburden** (*n*) rock or other material of no value that lies over a deposit of useful material.

**quarry** (*n*) a place where rock is dug out in the open. **quarry** (*v*), **quarried** (*adj*).

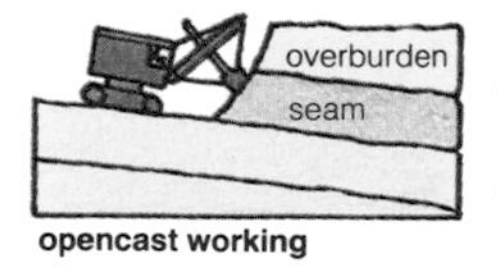


opencast working

**hydrology** (*n*) the study of water as it occurs on the Earth; in streams, as runoff (↓), from springs, etc.

**hydrogeology** (*n*) the geology of water supplies that are obtained from under the ground.

**precipitation** (*n*) water falling as liquid (rain, dew) or solid (snow, hail, frost, etc.) on the surface of the Earth from the atmosphere.

**runoff** (*n*) the water falling on the Earth's surface that reaches the streams and rivers.

**catchment area** the area from which the rainwater or other precipitation (↑) enters a particular river or stream. Outside the catchment area the water flows in another direction.

**groundwater** (*n*) water that is present in the pore spaces (p.84) and other spaces in the rocks below the Earth's surface.

**water table** the upper surface of the groundwater (↑); the surface below which the pore spaces (p.84) of the rocks are filled with water.

**spring** (*n*) water from under the ground coming out at the surface of the Earth. Springs occur where the water table (↑) meets the ground surface or where water under pressure reaches the surface.

**aquifer** (*n*) a stratum (p.80) of rock below the Earth's surface that holds water and through which water can move.

**artesian** (*adj*) refers to an aquifer (↑) with impermeable (↓) beds above it and in which water is under a high enough pressure for it to rise above the aquifer.

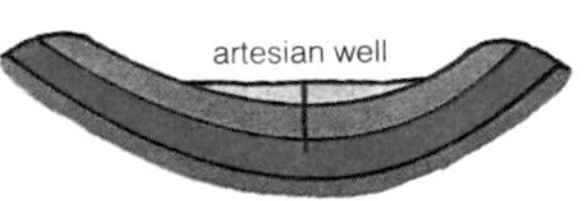

**artesian basin**

**meteoric water** water at and below the surface of the Earth that has come from the atmosphere, i.e. from precipitation (↑).

**juvenile water** water in the Earth's crust that has come from magma (p.62).

**connate water** water in a sedimentary rock (p.80) that is believed to have been trapped in the sediment at the time it was formed.

**permeability** (*n*) in hydrology (↑), permeability is a measure of the ease with which liquids and gases can pass through a rock. **permeable** (*adj*).

**impermeable** (*adj*) not permeable (↑).

**specimen** (*n*) a thing or part of a thing – e.g. a rock or a fossil (p.98) – that is taken as an example.
**hand specimen** a piece of a rock of a size that is suited to examination in the laboratory (↓).
**laboratory** (*n*) a room or rooms set aside for scientific work.
**lens** (*n*) a piece of glass with curved surfaces that can be used to make an object (e.g. a rock specimen) appear larger.
**microscope** (*n*) an instrument that uses lenses (↑) to give a view of a small object in which its size appears to be greatly increased.

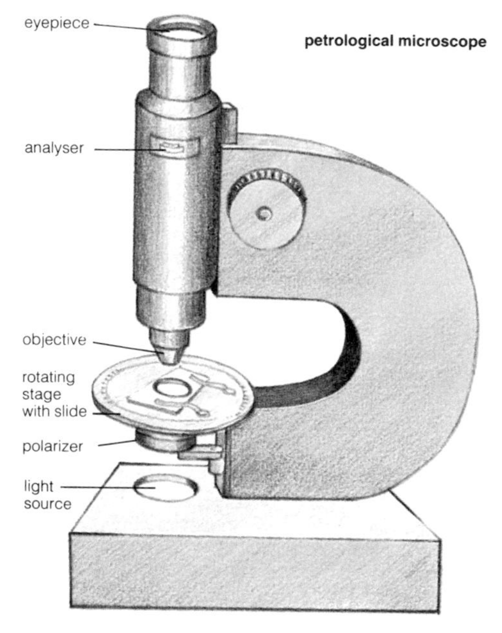


**petrological microscope** a microscope (↑) with special fittings for studying rocks and minerals by using polarized light (p.158).
**thin section** a very thin piece of a rock or mineral fixed to a piece of glass so that it can be viewed under a microscope (↑).

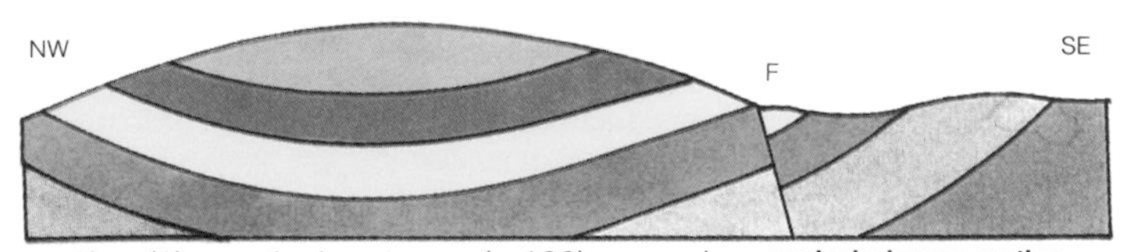


geological cross-section

**section** (1) a vertical exposure (p.122) or a series of exposures of rocks, e.g. in a cliff; (2) also **cross-section**: a drawing made to show the geological structure (p.122) along a chosen line; (3) also **cross-section**: a cut made across a specimen (p.147) to show what is inside it; (4) a thin section (p.147).

**cross-section** (*n*) *see* **section** (2), (3) (↑).

***in situ*** (*Latin*) in place, e.g. of a rock or fossil that has not been moved from the place where it was formed.

**contact** (*n*) the surface at which two different kinds of rock come together; especially between an igneous rock (p.62) and the country-rock (p.65).

**clinometer** (*n*) a simple instrument for measuring angles from the horizontal, such as angles of dip (p.123).

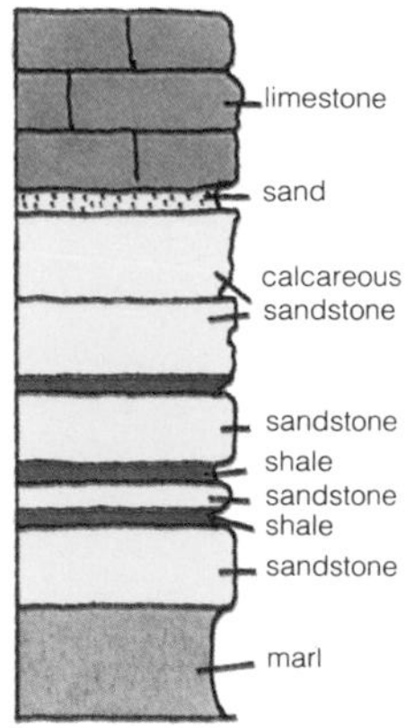


geological section

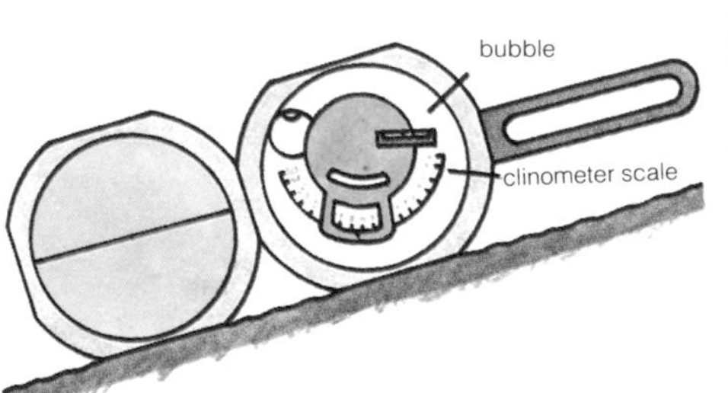


**Burton pocket transit**
when it is used as a clinometer the bubble is centred by moving a lever (not shown here) on the back of the case; the angle of dip (p.123) is then read off from the scale.

**prismatic compass** a hand compass fitted with sights so that the user can measure angles to chosen points.

**field work, field geology** geological work done in the open air outside the laboratory (p.147).

**survey** (*v*) to make measurements of an area of land and draw a map of it.

**geological survey** the work of studying the geology of part of the Earth's crust (p.9) and drawing a geological map of it.

meteorite

**meteorite** (*n*) a small solid body from space that has fallen on the Earth's surface.
**stony meteorite** a meteorite made up chiefly of silicate minerals (pp.16, 53) such as olivine (p.58), pyroxenes (p.57), and feldspars (p.56).
**stone** (*n*) = stony meteorite (↑).
**aerolite** (*n*) = stony meteorite (↑).
**chondrite** (*n*) a stony meteorite (↑) containing *chondrules*: small round masses about 1 mm in diameter made up of olivine (p.58) and pyroxene (p.57). *See also* **achondrite** (↓).
**achondrite** (*n*) a stony meteorite (↑) that does not contain the *chondrules* that are found in chondrites (↑).
**siderite** (*n*) a meteorite (↑) consisting of metals: about 90% of iron (Fe) and 6 to 9% of nickel (Ni).
**iron** (*n*) = siderite (↑).
**siderolite** (*n*) a meteorite (↑) containing more or less equal amounts of metals and silicates (p.16).
**stony-iron** (*n*) = siderolite (↑).
**tektite** (*n*) a rounded, flat, glassy meteorite (↑).

**lunar** (*adj*) of the moon.
**mare** (*n*) (*maria*) one of the large and generally flat areas of the moon which appear dark in colour as seen from the Earth. These areas are made up of mafic (p.75) and ultramafic (p.75) rocks.
**lunar highlands** areas of the Moon's surface that are higher than the maria (↑), with many craters (↓). The lunar highlands appear bright as seen from the Earth.
**terrae** (*n.pl.*) = lunar highlands (↑).
**crater** (*n*) a circular hollow with steep slopes formed either by volcanic action (p.68) or by the fall of a meteor (p.149).
**mascon** (*n*) an area on the Moon where the density (p.154) of the rocks below the surface is especially high.
**rille** (*n*) one of the long, narrow valleys on the surface of the Moon. Rilles are up to several hundred kilometres long and one to two kilometres wide. The walls of the valley are steep and its bottom is flat. A rille may be straight (a *normal rille*) or winding (a *sinuous rille*).
**mare ridge** a long, narrow hill in a lunar mare (↑). Mare ridges are up to a few hundred kilometres long and several tens of metres high.
**wrinkle ridge** = mare ridge (↑).
**lunar regolith** a thin layer of grey material on the surface of the Moon. It consists of loose or partly cemented (p.84) fragments ranging from very fine dust to large blocks.
**lunar soil** = lunar regolith (↑).

# Additional definitions

The following list gives definitions of words that are not in the basic world list (the defining vocabulary) that is used for most of the definitions in the main part of this dictionary. These additional words are needed to explain some of the geological terms that are included in the dictionary. Some of them are words in everyday use; others are scientific terms that are only indirectly related to geology.

For ease of reference the words in this appendix are listed in alphabetical order.

**acceleration** (*n*) the increase in velocity (p.153) per unit time, i.e. the increase in velocity divided by the time taken to increase it. If the velocity of a motor-car increases from 20 metres per second (m/s) to 30 m/s in 5 seconds, then the acceleration = (increase in velocity) ÷ (time) = (30 – 20) m/s ÷ 5 s = 10 m/s ÷ 5 s = 2 $m/s^2$ (metres per second per second) **accelerate** (*v*), **accelerated, accelerating** (*adj*).

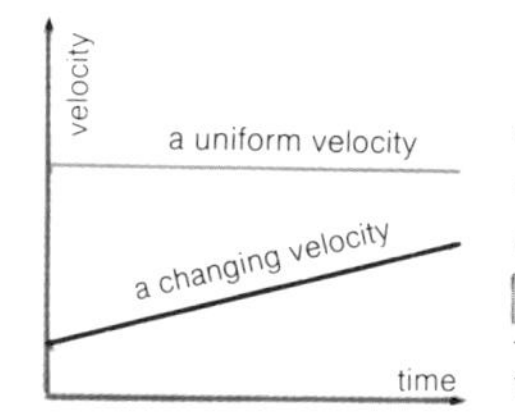


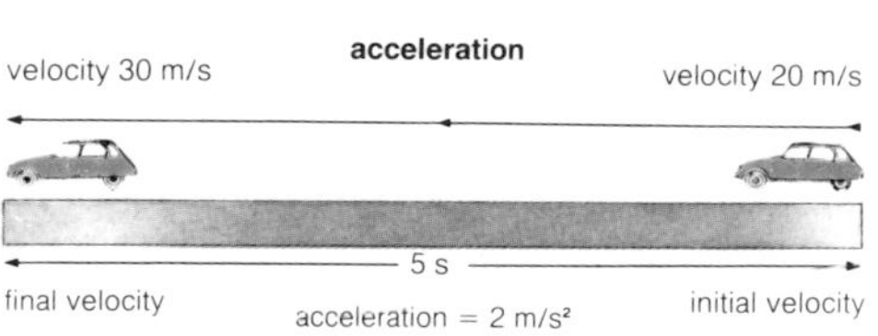


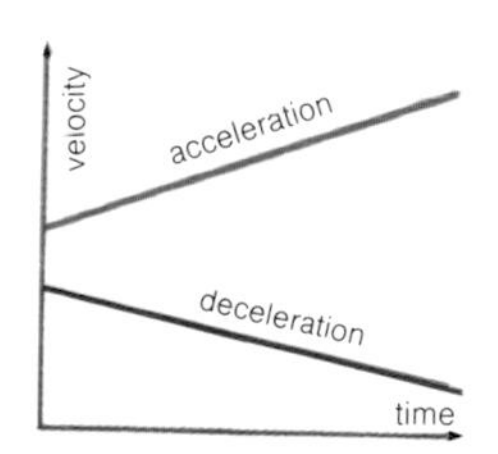


**accumulate** (*v*) to build up in one place, to heap up in a mass. **accumulation** (*n*).

**aggregate** (*v*) to gather together into one whole; to mass together. **aggregate** (*n*).

**alter** (*v*) to make something different in some way without changing the thing itself. **alteration** (*n*).

**alternate** (*adj*) (of two things) arranged or coming one after the other by turns. **alternately** (*adv*), **alternation** (*n*).

**atom** (*n*) the smallest particle of an element (p.15) that has the properties of that element and takes part in chemical reactions (p.17). **atomic** (*adj*).

**atomic number** the number of protons (p.151) in the nucleus (p.150) of an atom (↑). It determines the chemical nature of the atom, i.e. to which element an atom belongs.

**atomic weight** the average weight of the atoms (↑) of an element in relation to the oxygen atom, which for this purpose is taken to be 16.

**average** (*n*) an average is the sum of variable quantities divided by the number of the quantities, e.g. the average of 10 m, 16 m, 8 m, 12 m, is: (10 m + 16 m + 8 m + 12 m) ÷ 4 = 46 m ÷ 4 = 11.5 m. **average** (*adj*).

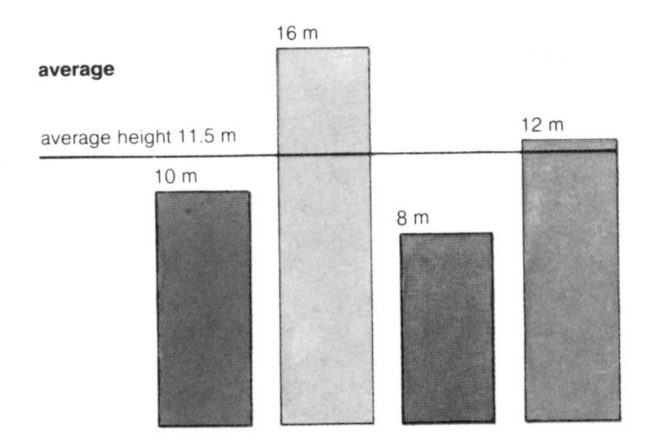


**balance** (*n*) a state in which two or more things are equal to each other in weight or some other respect so that there is no tendency for them to move up or down. **balanced** (*adj*).

**base** (*n*) a substance which reacts with an acid to form a salt and water only, generally an oxide or a hydroxide of a metal. **basic** (*adj*).

**boundary** (*n*) the edge or border of a thing.

**cartilage** (*n*) a material in the body of a vertebrate (p.109) that holds other parts of the body together (e.g. bones). Some animals, e.g. sharks, have cartilage instead of bone in their skeletons (p.159).

**cell** (*n*) the smallest part of a plant or animal. The simplest living things consist of only one cell; others, e.g. mammals (p.109), contain many millions of cells. A cell has the ability to take in chemical substances and use them to make the various substances it needs in order to live. **cellular** (*adj*).

**channel** (*n*) a hollow bed in which water runs; a long narrow hollow.

**charge** (*n*) an electric charge cannot be explained but a body that carries a charge will be drawn towards another body carrying a charge of opposite sign (i.e. a body with a positive charge will be drawn towards a body with a negative charge); and bodies with charges of the same sign (both positive or both negative) will tend to be pushed away from each other.

**chemistry** (*n*) the study of the elements and their compounds (p.15), their nature and the ways in which they act upon each other. **chemical** (*adj*).

**classify** (*v*) to arrange in classes. **classification** (*n*), **classified** (*adj*).

**coarse** (*adj*) made up of large pieces or particles, etc.; the opposite of 'fine'.

**combine** (*v*) to join together. **combined** (*adj*).

**complex** (*adj*) made up of many parts; not simple. **complexity** (*n*).

**compose** (*v*) to form by being together, e.g. minerals may compose a rock. **composed** (*adj*).

**concave** (*adj*) curved inwards; hollow. See also **convex** (p.154).

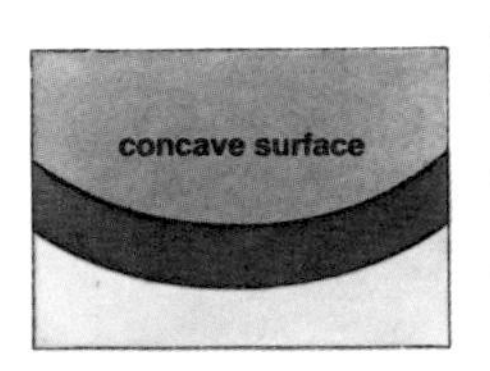


**concentric** (*adj*) (of circles) having the same centre. **concentrically** (*adv*).

**conductor** (*n*) a material through which heat or an electric current can flow. All metals are good conductors of heat and electricity. **conduct** (*v*).

**cone** (*n*) a solid figure produced by a straight line passing through a fixed point and a circle. **conical** (*adj*).

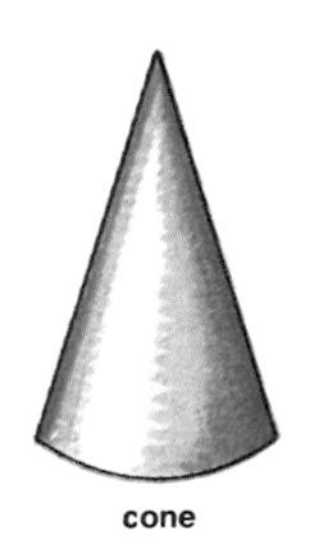
cone

**continent** (*n*) one of the larger unbroken land masses of the Earth's surface, e.g. Africa.

**contour** (*n*) a line joining points on a map or diagram that have the same value; (usually) points on a map that have the same height.

**contract** (*v*) to decrease in length, area, or volume of a solid, or in volume of a fluid. Contraction can be caused by a fall in temperature. **contraction** (*n*).

**convection** (*n*) the transfer of heat in a fluid by the rising of hotter fluid and the sinking of colder fluid to take the place of the hotter fluid. A **convection current** is formed by the movement of the fluid.

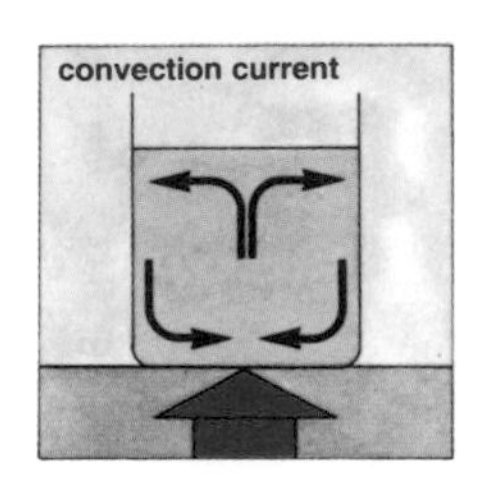


**convex** (*adj*) of round shape on the outside; the opposite of **concave** (p.153).

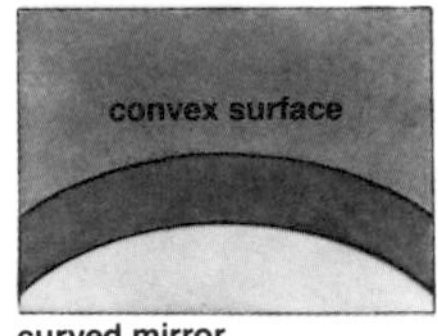

curved mirror

**correspond (to)** (*v*) to be similar to in some character of working. **corresponding** (*adj*).

**decompose** (*v*) to break down something into the parts of which it is made. **decomposition** (*n*), **decomposed** (*adj*).

**density** (*n*) mass per unit volume, i.e. the mass of a material divided by its volume, e.g. 128.2 $cm^3$ of iron has a mass of 1 kg: density of iron = 7.8 $g/cm^3$ = 7800 $kg/m^3$ (kilograms per cubic metre). Each material has a particular density, e.g. the density of water is 1 $g/cm^3$. Density is important in the identification (p.93) of materials. **dense** (*adj*).

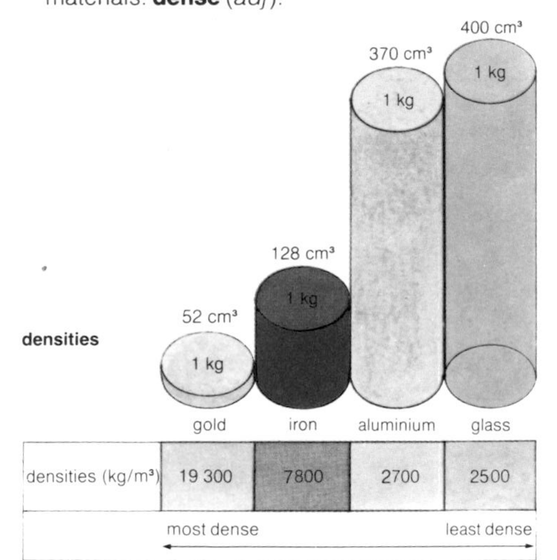

densities

| densities ($kg/m^3$) | 19 300 | 7800 | 2700 | 2500 |
|---|---|---|---|---|
| | most dense | | | least dense |

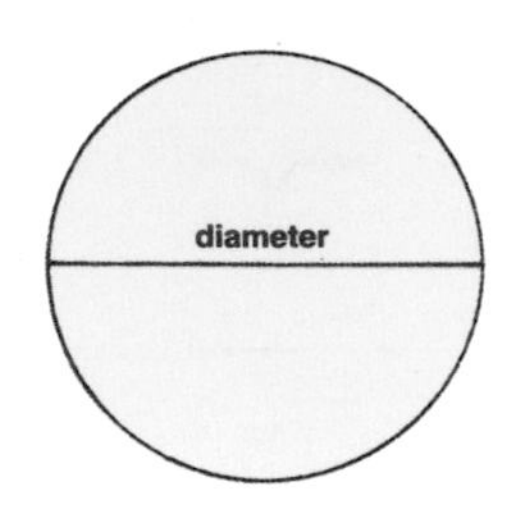


**diagram** (*n*) a drawing or figure that is intended to explain something or to show the relationships between two or more things.

**diameter** (*n*) a straight line passing from side to side through the centre of a circle or a sphere, or the length of that line.

**dimension** (*n*) the dimensions of a solid are its length, its breadth, and its height. Liquids and gases do not have dimensions. A solid has three dimensions: a flat surface has two dimensions (length and breadth); and a line has only one dimension (length). **in three dimensions** in the three directions of space. **dimensional** (*adj*).

**dissolve** (*v*) to put into solution (↑). **dissolved** (*adj*).

**distillation** (*n*) a physical process for separating liquids from mixtures. The mixture is heated and the liquid is turned to gas; the gas is then cooled in a tube, where it turns to liquid and can be collected. Liquids with different boiling points can be separated by this process. **distil** (*v*).

**distribute** (*v*) to spread out or scatter in space or over a surface. **distribution** (*n*).

**division** (*n*) the act of dividing; the state of being divided.

**downwards** (*adv*) from higher to lower.

**drill** (*v*) to make a hole through or into something, e.g. to drill a borehole (p.139) into the Earth.

**electron** (*n*) a very small particle with a mass about 1/1840 that of a hydrogen atom (↑) and a very small negative electrical charge (p.153).

***en echelon*** (*adj, French*) used to describe structures (e.g. folds) that are parallel to each other but are not opposite to each other.

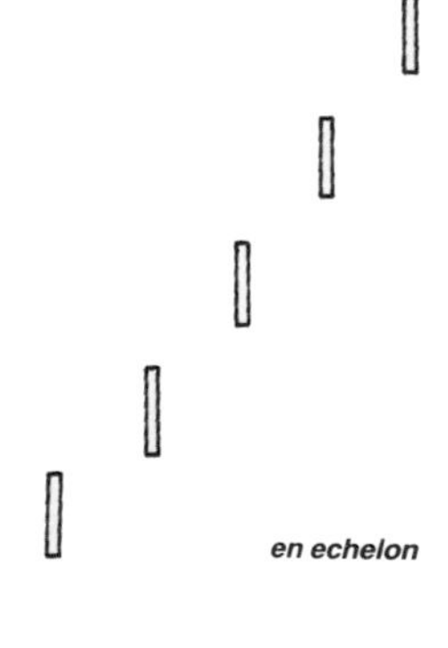


**energy** (*n*) the ability to do work. There are different forms of energy: potential energy (stored energy); kinetic energy (energy from motion); heat energy; light energy; electrical energy; chemical energy; nuclear energy. One form of energy can be transformed into another.

**equilibrium (equilibria)** (*n*) a state of balance (p.146) between opposing forces or effects. Forces that balance each other are *in equilibrium*. An object is equilibrium if the forces acting upon it are in equilibrium.

**equivalent** (*n*) equal in value, power, meaning, etc., to something else.

**evaporation** (*n*) the change of a liquid to a vapour at a temperature below, or at, its boiling point, e.g. the evaporation of rain water without the water boiling; the evaporation of a salt solution (p.159) leaving the salt. **evaporate** (*v*).

**expand** (*v*) to increase in length, area, or volume of a solid, or in volume of a liquid or gas. **expansion** (*n*).

**external** (*adj*) lying outside.

**fine** (*adj*) made up of small pieces or particles, etc. The opposite of 'coarse' (p.147).

**force** (*n*) a push or pull that causes: (1) an acceleration, or (2) a change in the shape of an object, or (3) a reaction (the opposite effect to an action, equal to it but in the opposite direction). A force can be measured by: (1) the amount it stretches a spring; (2) the acceleration it gives to a mass.

**formula** (*n*) chemical symbols written together to show the atoms in a molecule of a compound or in an ion; e.g. (a) the formula MgO stands for a molecule of magnesium oxide and shows that it is composed of one atom of magnesium combined with one atom of oxygen; (b) the formula $NO_3^-$ stands for a nitrate ion. The formula gives the composition of a substance.

**fragment** (*n*) a piece broken off. **fragmented** (*adj*).

**frequency** (*n*) the number of times an event is regularly repeated in unit time, e.g. the number of vibrations in 1 second. **frequent** (*adj*).

**gas** (*n*) a state of matter (p.159) in which the molecules (p.15) are free to move about, there being no forces to hold them together. A gas has no definite volume and no shape. **gaseous** (*adj*).

**gene** (*n*) a short length of a chromosome (a thread-like body in the part of a cell that controls the activities of a cell) which controls a characteristic of a living thing. A gene can be passed on from parent to the next generation.

**geography** (*n*) the study of the Earth's surface; its physical features, etc. **geographical** (*adj*).

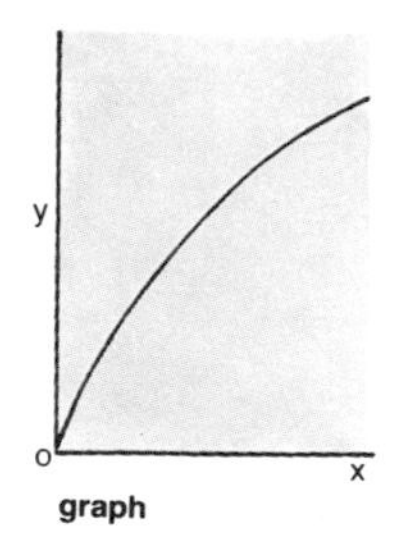

graph

**graph** (*n*) a diagram that shows the variation of a quantity in relation to another quantity, e.g. a graph of density against temperature for a given substance. **graphical** (*adj*).

**horizontal** (*adj*) level; parallel with the horizon (the line where the Earth and sky appear to meet).

**hydration** (*n*) a chemical process in which a compound (p.15) absorbs and combines with water. **hydrated** (*adj*).

**hydrous** (*adj*) containing water.

**intense** (*adj*) to a high degree; very strong, violent.

**landscape** (*n*) the scene presented by a piece of country.

**layer** (*n*) a thickness of a substance spread over a surface, especially one of a series. **layered** (*adj*).

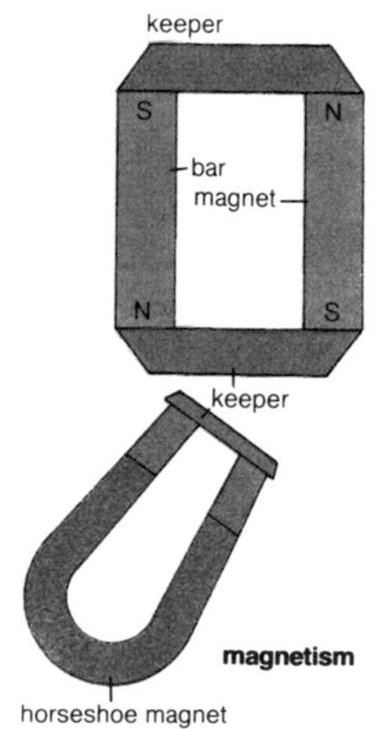

magnetism

**magnet** (*n*) a solid object that attracts iron and attracts or pushes away (repels) other magnets. When free to turn, it points in a north – south direction. Magnets possess the property of magnetism (*n*). **magnetic** (*adj*), **magnetize** (*v*).

**magnetic field** the space round a magnet of an electric current in which a magnetic material experiences a magnetic force of attraction, or a magnet sets in the direction of the magnetic force from the magnet.

**margin** (*n*) the part of a surface that lies just inside its border; an edge, a border line. **marginal** (*adj*).

**marine** (*adj*) (1) of the seas and oceans; (2) inhabiting the sea, found in or formed by the sea.

**medium** (*adj*) between the highest and lowest levels; in the middle.

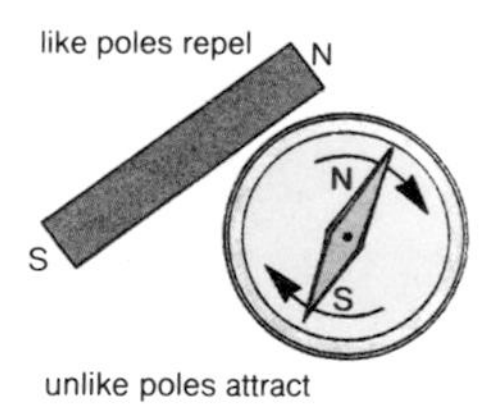


**method** (*n*) a means of doing something.

**nucleus** (*atomic*) (*n*) the central part of an atom. It consists of one or more protons and has a positive charge (p.146) which is almost exactly equal to the total charge on the electrons that surround it in the atom. Nearly the whole of the mass of the atom is in the nucleus. **nuclear** (*adj*).

**occur** (*v*) to be found, to happen.

**optical interference** under certain conditions two sets of light waves can act on each other, producing alternate light and dark bands (if the light is of one wavelength) or colours like those of the rainbow (if white light is used). (The colours seen in thin films of oil on water are an example.) Interference can occur in an anisotropic (p.47) mineral, and the colours that result can be used in identifying the mineral.

**origin** (*n*) the first state of something; the first place from which something has come. **original** (*adj*).

**oscillate** (*v*) to move regularly to and fro or up and down. **oscillation** (*n*), **oscillating** (*adj*).

**parallel** (*adj*) (of two lines or planes) running in the same direction and everywhere at the same distance from each other; never meeting.

**particle** (*n*) a small or very small piece of something.

**physics** (*n*) the study of the properties of matter (except for their chemical properties) and of energy (p.148). **physical** (*adj*).

**plan** (*n*) a view or a drawing of something as seen from above.

**plane** (*n*) a surface that is flat. In geology the word plane is also used for surfaces that are not quite flat. **planar** (*adj*).

**plastic** (*adj*) able to be shaped into any form and of remaining in that shape. Clays (p.88) are plastic.

**polarized light** in ordinary light, vibrations take place in all possible directions at right angles to the path of the light beam. In plane polarized light the vibrations take place in only one plane. Polarized light is used in the petrological microscope for the identification of minerals.

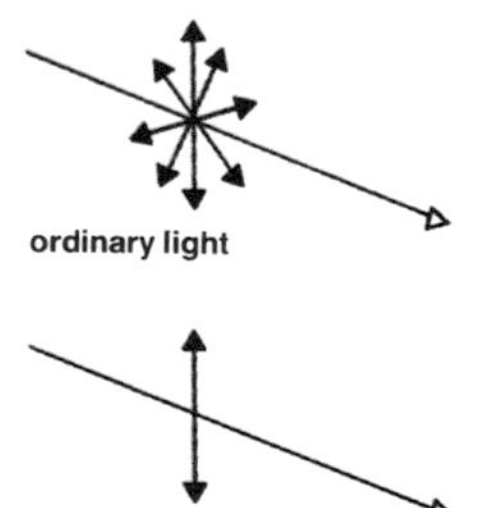


**plane polarized light**

**position** (*n*) the place in which something is found.

**process** (*n*) a set of actions or events taking place one after the other.

**property** (*n*) something that a thing always, or usually, has or shows; a characteristic.

**proportion** (*n*) the relationship of two or more things to each other in their sizes, quantities, or numbers, etc. **proportional** (*adj*).

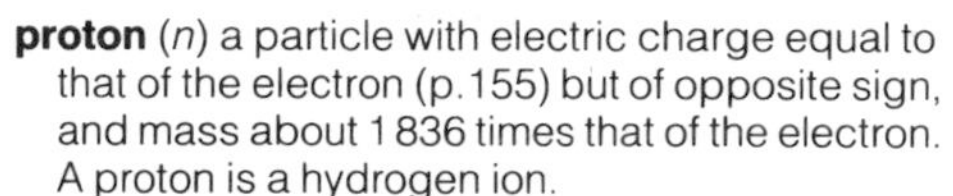

**proton** (*n*) a particle with electric charge equal to that of the electron (p.155) but of opposite sign, and mass about 1 836 times that of the electron. A proton is a hydrogen ion.

**ratio** (*n*) the relationship between two quantities expressed in numbers, e.g. if a mixture contains 2 parts of A to 5 parts of B then the ratio of A to B in it is 2:5.

**refer** (*v*) to relate; to direct attention; to turn for facts or other knowledge. **reference** (*n*).

**refraction**

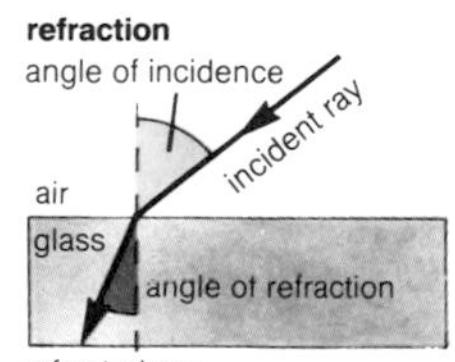


**refractive index** a measure of the ability of a material to bend a beam of light. For a particular material it is equal to the speed of light in a vacuum divided by the speed of light in the material. It is also measured by: {sine (angle of incidence)} ÷ {sine (angle of refraction)}: see diagram.

**replace** (*v*) to take the place of. **replacement** (*n*), **replaceable** (*adj*).

**ridge** (*n*) a long narrow area of high ground; the edge where two slopes meet at the top.

**rigid** (*adj*) describes a solid that does not change in shape when a force acts upon it.

**series** (*n*) a set of things in a line or following one after the other, and having something in common.

**skeleton** (*n*) the frame of bone or other hard material that supports or contains an animal. **skeletal** (*adj*).

**solution** (*n*) the change of matter from the solid or gaseous state to the liquid state by putting it in a liquid.

**stable** (*adj*) not easily changed or moved.

**states of matter** all materials are solids, liquids, or gases. These are the three states of matter.

**subsurface** (*adj*) below the surface, e.g. subsurface geology, the geology of the region below the Earth's surface.

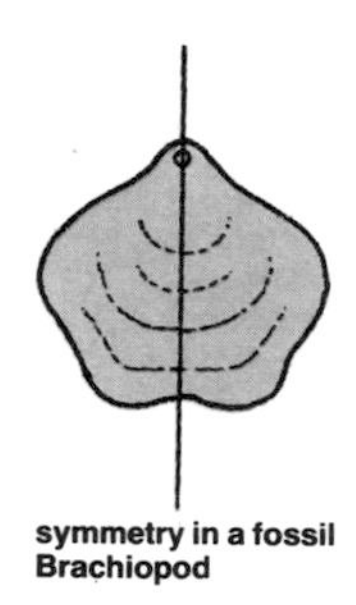

**symmetry in a fossil Brachiopod**

**symmetry** (*n*) having parts that correspond (p.155) on either side of a plane, a straight line, or a point. Parts on opposite sides of the plane, line, or point are similar in size, shape, and position. In crystallography the word 'symmetry' is used with a special meaning: see p.42. **symmetrical** (*adj*).

**temperature** (*n*) a measure, using a scale, of how hot, or how cold, an object, an organism (p.98), or the atmosphere is.
**term** (*n*) a word used with a special meaning, as in science. Thus 'rock' used as a term in geology has a different meaning from the meaning it has in ordinary speech.
**theory** (*n*) a supposition that is put forward to explain a body of facts; e.g. the theory of isostacy (p.11); the theory of plate tectonics (p.134). **theoretical** (*adj*).
**tissue** (*n*) (in living things) a mass of cells (p.146) and the material between them that together have the same purpose.
**unit** (*n*) a thing or a group of things that can be regarded as the smallest part into which something larger can be divided; a quantity chosen as a standard measurement for other quantities.
**upward** (*adv*) moving towards a higher place.
**variation** (*n*) varying (↓) from a normal or earlier state or amount or standard.
**vary** (*v*) to change or make (or become) different. **varied** (*adj*).
**velocity** (*n*) speed in a given direction. A motor-car travelling along a straight road with a speed of 70 km/h has a velocity of 70 km/h. A motor-car travelling at 70 km/h round a bend in a road has a velocity that is changing all the time, because it is not travelling in a straight line.

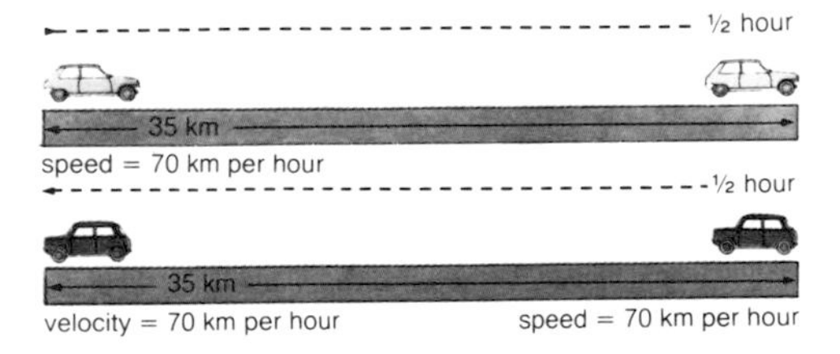


**vertical** (*adj*) upright; at 90° to the plane of the horizon (the line where the Earth and the sky appear to meet).
**vibrate** (*v*) to move to and fro, especially rapidly. **vibration** (*n*), **vibrating** (*adj*).

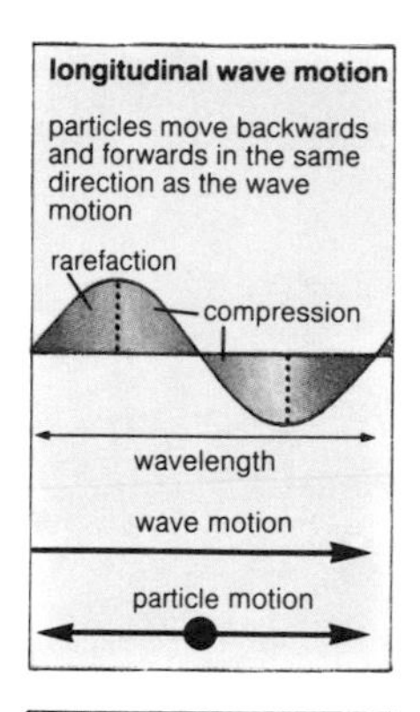


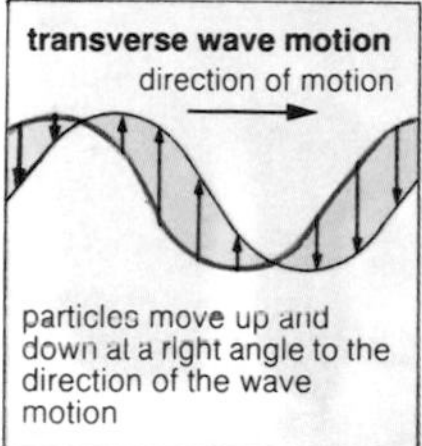


**volume** (*n*) the amount of space filled by a body or a substance. Volume is usually measured in cubic metres ($m^3$) or a related unit ($mm^3$, $km^3$ for example).

**wave motion** the sending of energy by a regular movement, in the form of a wave (see diagram). When a wave passes through a material, each particle moves about a central point to produce the wave motion. Waves are of two kinds: transverse, in which the particles move at 90° to the direction in which the wave is travelling; and longitudinal, in which the particles move backwards and forwards in the same direction as the wave is travelling.

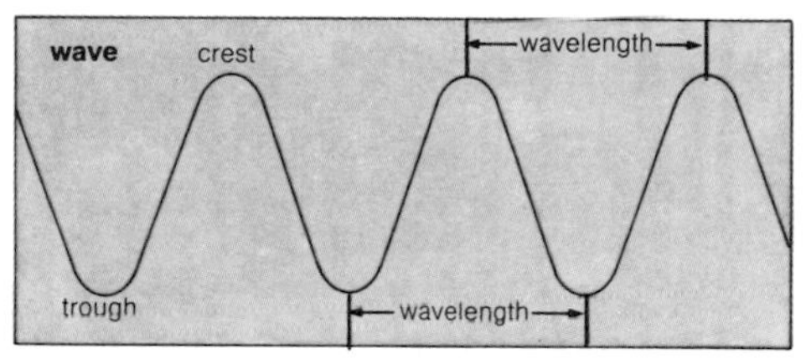


**wavelength** (*n*) the distance between a point in a wave and the next point at the corresponding place moving in the same direction; i.e. the distance between one crest (see diagram) and the next crest or between one trough and the next.

**zone** (*n*) in the general sense, a long narrow area shaped like a band, especially one that stretches like a ring round something. **zonal** (*adj*). 'Zone' is used in a special sense in palaeontology and stratigraphy: see pp.111, 117.

# Common abbreviations in geology

| | |
|---|---|
| aff. | affinity |
| anh | anhydrous |
| approx. | approximate(ly) |
| aq. | aqueous |
| av. | average |
| b.p. | boiling point |
| BP | before the present |
| *c.* | circa (about) |
| corr. | corrected |
| cryst. | crystalline |
| cy | clay |
| exptl | experimental |
| f.p. | freezing point |
| gen. | genus |
| H | hardness (Mohs' scale) |
| L. | lower |
| liq. | liquid |
| lst | limestone |
| M. | middle |
| max. | maximum |
| min. | minimum |
| m.p. | melting point |
| pH | hydrogen ion concentration |
| ppm | parts per million |
| ppt | precipitate |
| R.I. | refractive index |
| s.l. | sea level |
| *s.l.* | *sensu lato* (in the broad sense) |
| sol. | soluble |
| soln | solution |
| sp., spp. | species (sing., plural) |
| s.s | *sensu stricto* (in the strict sense) |
| sst | sandstone |
| s.t.p. | standard temperature and pressure |
| temp. | temperature |
| U. | upper |
| vol. | volume |
| v.p. | vapour pressure |
| wt % | weight per cent |

# International System of Units (SI)

PREFIXES

| PREFIX | FACTOR | SIGN | PREFIX | FACTOR | SIGN |
|---|---|---|---|---|---|
| milli- | $\times 10^{-3}$ | m | kilo- | $\times 10^{3}$ | k |
| micro- | $\times 10^{-6}$ | $\mu$ | mega- | $\times 10^{6}$ | M |
| nano- | $\times 10^{-9}$ | n | giga- | $\times 10^{9}$ | G |
| pico- | $\times 10^{-12}$ | p | tera- | $\times 10^{12}$ | T |

BASIC UNITS

| UNIT | SYMBOL | MEASUREMENT |
|---|---|---|
| metre | m | length |
| kilogramme | kg | mass |
| second | s | time |
| ampere | A | electric current |
| kelvin | K | temperature |
| mole | mol | amount of substance |

DERIVED UNITS

| UNIT | SYMBOL | MEASUREMENT |
|---|---|---|
| newton | N | force |
| joule | J | energy, work |
| hertz | Hz | frequency |
| pascal | Pa | pressure |
| coulomb | C | quantity of electric charge |
| volt | V | electrical potential |
| ohm | $\Omega$ | electrical resistance |

SOME MULTIPLES OF SI UNITS HAVING SPECIAL NAMES

| UNIT | SYMBOL | DEFINITION | MEASUREMENT |
|---|---|---|---|
| ångstrom | Å | $10^{-10}$ m = $10^{-1}$ nm | length |
| micron | $\mu$m | $10^{-6}$ m | length |
| litre | l | $10^{-3}$ m$^3$ = dm$^3$ | volume |
| tonne | t | $10^3$ kg | mass |
| dyne | dyn | $10^{-5}$ N | force |
| bar | bar | $10^5$ Pa | pressure |

SOME NON-SI UNITS

| UNIT | SYMBOL | DEFINITION | MEASUREMENT |
|---|---|---|---|
| atm | atm | 101325 Pa, 1.01325 bar | pressure |
| degree Celsius | °C | K ($t_c = t_k - 273$) | temperature |
| million years | Ma, m.y. | $10^6$ years | time |
| billion (US) years | Ga | $10^9$ years | time |

# Understanding scientific words

New words can be made by adding **prefixes** or **suffixes** to a shorter word. Prefixes are put at the front of the shorter word and suffixes are put at the back of the word. Words can also be broken into parts, each of which can have a meaning, but cannot be used alone.

*(i)* correct → *in*correct (adding a prefix)
correct → correct*ness* (adding a suffix)
correct → *in*correct*ness* (adding a prefix and a suffix)

*(ii)* **isomorphism** is broken into iso-morph-ism
**iso-** is a prefix which means 'identical in structure'
**morph** is a word part which means 'form or shape'
**-ism** is a suffix which means 'a condition'

Hence *isomorphism* means the condition of having identical forms or shapes; it describes the condition of two crystalline substances.

Prefixes describing numbers or quantities are taken from Greek or Latin words. The following table shows the common prefixes from these two languages. Prefixes, suffixes, and word parts are listed alphabetically in separate sections after the table.

| | GREEK PREFIX | LATIN PREFIX | PREFIX | MEANING | |
|---|---|---|---|---|---|
| 1 | mono- | uni- | hemi- | half | Gr |
| 2 | di- | bi- | semi- | half | L |
| 3 | tri- | ter- | poly- | many | Gr |
| 4 | tetra- | quad- | multi- | many | L |
| 5 | penta- | quinq- | omni- | all | L |
| 6 | hexa- | sex- | dupli- | twice | L |
| 7 | hepla- | sept- | tripli- | three times | Gr |
| 8 | octo- | oct- | hypo- | less, under | Gr |
| 9 | nona- | novem- | hyper | more, over | Gr |
| 10 | deca- | deci- | sub- | under | L |
| 100 | hecta- | centi- | super- | over | L |
| 1000 | kilo- | milli- | iso- | same, equal, identical | Gr |

## PREFIXES

**a-** without, lacking, lacking in, e.g. *a*morphous, being without shape; *a*symmetrical, without symmetry, or lacking in symmetry.

**allo-** different, or different kinds, e.g. *allo*tropy, the existence of an element in two or more different forms.

**amphi-** on both sides, e.g. *ampho*teric, having the nature of both an acid and a base.

**an-** the same prefix as **a-**, used in front of words beginning with a vowel, or the letter *h*, e.g. *an*isotropic, not having the same properties in all directions; *an*hydrous, being without, or lacking, water, in a crystal.

**anti-** opposite in direction, or in position, e.g. *anti*catalyst, a catalyst which slows down a chemical reaction, i.e. works in the opposite direction to a catalyst.

**auto-** caused by itself, e.g. *auto*xidation, reaction of a substance with atmospheric oxygen at room temperature, the substance oxidizes itself; *auto*catalysis, a chemical reaction in which the products act as catalysts for the reaction.

***cis-*** on the same side, e.g. *cis*-compound, an isomer in which two like groups are on the same side of the double bond in the compound. See *trans-*.

**co-** acting together, with, e.g. *co*hesion, the force holding two or more objects together.

**counter-** acting against, acting in the opposite direction, e.g. *counter*act, to act against, such as a mild alkali counteracts the effect of acid on skin; *counter*clockwise, turning in the opposite direction to the hands of a clock.

**de-** opposite action, e.g. *de*compression, the lessening of a pressure, it is the opposite action to compression; *de*activate, to make less active, it is the opposite of activate.

**dia-** through, across, e.g. *dia*meter, the line going across a circle, through the centre.

**dis-** opposite action, e.g. *dis*charge, to take an electric charge away from a charged body, the opposite of charge; *dis*connect, to break, or open, a connection, the opposite of connect.

**equi-** having the same number, equal, e.g. *equi*molecular, having the same number of molecules; *equi*librium, the condition of two rates of reaction being equal and opposite, so that there is no further change in a reversible reaction.

**im-** the opposite, not. (Used with words beginning with b, m, p.) For example, *im*perfect, not perfect, the opposite of perfect; *im*permeable, not permeable.

**in-** the opposite, not. (Used with all words other than those beginning with b, m, p.) For example, *in*active, the opposite of active; *in*adequate, not adequate.

**infra-** below, e.g. *infra*molecular, having a size smaller than a molecule, so the size is below molecular size.

**inter-** between, among, e.g. *inter*face, a common surface between two liquids or two solids; *inter*stice, a narrow space between two solid objects.

**macro-** great, large, e.g. *macro*molecule, a large molecule composed of many smaller molecules, as in a polymer.

**micro-** small, especially if too small to be seen by the human eye alone, e.g. *micro*balance, a balance used for measuring masses of less than 1 mg; *micro*analysis, analysis using very small amounts of substances.

**non-** not, e.g. *non*-electrolyte, a substance which is not an electrolyte; *non*-ferrous, any metal other than iron.

**ortho-** straight, right-angled, upright, e.g. *ortho*gonal, with parts at right-angles; *ortho*rhombic, a crystal system with three unequal axes at right-angles.

**pan-** all, complete, every, e.g. *pan*chromatic, covering all wavelengths of light in the spectrum.

**para-** at the side of, by, e.g. *para*casein, an insoluble form of casein, formed when soluble casein coagulates.

**pseudo-** has the same appearance, but is false, e.g. *pseudo*alum, a substance which has the appearance of an alum, but is not an alum.

**re-** again, e.g. *re*activate, to make something activated again; *re*crystallize, to crystallize again.

**syn, sym-** joined together, united, e.g. *syn*thesis, combining elements or compounds to make new compounds.

***trans-*** across, on the opposite side of, e.g. *trans*-compound, an isomer in which two like groups are on opposite sides of the double bond in the compound. See *cis-*.

**ultra-** beyond, e.g. *ultra*filter, a filter which has holes so small it filters out colloids; it thus has uses beyond those of the ordinary filter.

**un-** not, the opposite, e.g. *un*saturated, means not saturated; *un*stable, means not stable; *un*paired, means not in a pair, and so by itself.

## SUFFIXES

**-able** forms an adjective which shows an action can possibly take place, e.g. change*able*, something which can change; transform*able*, something which it is possible to transform.

**-al** of, or to do with; forms a general adjective, e.g. experiment*al*, of, or to do with, experiment; fraction*al*, of, or to do with, fractions; therm*al*, of, or to do with, heat.

**-ed** forms the past participle of a verb, can be used as an adjective; it shows an action under the control of an experimenter, e.g. vari*ed*, describes a quantity changed by an experimenter; dehydrat*ed*, describes a substance from which water has been removed under the control of an observer.

**-er (-or)** forms a noun from a verb and describes an agent, e.g. mix*er*, a device which mixes; desiccat*or*, a device that desiccates; generat*or*, a device that generates a gas.

**-gram** forms a noun describing a record which is written or drawn, e.g. chromato*gram*, the recorded result from an experiment on chromatography; tele*gram*, the written message recorded by telegraph.

**-graph** forms a noun describing an instrument or device that records variation in a quantity, or other information, e.g. thermo*graph*, a kind of thermometer which records changes of temperature over a period of time; tele*graph*, a device which records information in words.

**-ic** of, or to do with; forms a general adjective, e.g. bas*ic*, of, or to do with, a base; cycl*ic*, of, or to do with, a cycle; ion*ic*, of, or to do with, ions.

**-ify** forms a verb which is causative in action, e.g. pur*ify*, to cause to become pure; solid*ify*, to cause to become solid.

**-ing** forms the present participle of a verb, can be used as an adjective; it shows an action not under the control of an experimenter, e.g. fluctuat*ing*, describes a quantity varying above and below an average value, which cannot be controlled by an observer; disintegrat*ing*, describes a radioactive substance undergoing disintegration, as the process cannot be controlled by an observer.

**-ity** forms a noun of a state or quality, e.g. pur*ity*, the quality or state of being pure; acid*ity*, the quality of being acid.

**-ive** forms an adjective by replacing *-ion* in nouns; the adjective describes an agent producing the effect described by the noun, e.g. inhibit*ion* → inhibit*ive*, describes an agent causing inhibition; oxidat*ion* → oxidat*ive*, describes a process causing oxidation; explos*ion* → explos*ive*, describes an agent causing an explosion.

**-ize** forms a verb which is causative in the formation of something, e.g. ion*ize*, to cause ions to be formed; polymer*ize*, to cause polymers to be formed.

**-lysis** forms a noun describing the action of breaking down into simpler parts, e.g. hydro*lysis*, the decomposition of a compound by the action of water; electro*lysis*, the decomposition of a substance by an electric current.

**-meter** forms a noun describing an instrument which measures quantitatively, e.g. thermo*meter*, an instrument which measures temperature accurately; volt*meter*, an instrument which measured electric potential in volts.

**-metry** forms a noun describing a particular science of accurate measurement, e.g. thermo*metry*, the science of measuring temperature; hydro*metry*, the science of measuring the density of liquids.

**-ness** forms an abstract noun of state or quality, e.g. sweet*ness*, the quality of being sweet; soft*ness*, the quality of being soft.

**-ous** forms an adjective showing possession, or describing a state, e.g. anhydr*ous*, being in the state of not possessing water, homolog*ous*, in the state of being a homologue; homogen*ous*, in the state of having the same properties throughout a substance.

**-philic** forms an adjective describing a liking for something, e.g. proto*philic*, describes a substance which accepts protons.

**-phobic** forms an adjective describing a dislike for something, e.g. lyo*phobic*, describes a colloid which does not go readily into solution.

**-scope** forms a noun describing an instrument which measures qualitatively, e.g. spectro*scope*, an instrument by which spectra can be observed qualitatively; hygro*scope*, an instrument which measures qualitatively the humidity of the atmosphere.

**-scopy** forms a noun describing the use of instruments for observation in science, e.g. micro*scopy*, the use of microscopes for scientific observation.

**-stat** forms a noun describing a device which keeps a quantity constant, e.g. hydro*stat*, a device which keeps water in a boiler at a constant level; thermo*stat*, a device which keeps a liquid, or an object, at a constant temperature.

**-tion** forms an abstract noun. With *-ation*, it forms a noun of action, e.g. pollu*tion*, the result of polluting; concentra*tion*, the degree to which a solution is concentrated; distill*ation*, the noun of action from distil; precipit*ation*, the noun of action from precipitate.

## WORD PARTS

**aqua** water, to do with water, e.g. *aqu*eous, a solution containing water; *aqua*ion, an ion with molecules of water associated with it.

**chrom** colour, to do with colour, e.g. pan*chrom*atic, all the colours, and hence all the wavelengths of the visible spectrum; *chrom*atography, the analysis of complex substances in which a coloured record of the analysis is produced.

**gen** to produce, e.g. homo*gen*ize, to make a mixture of solid and liquid substances into a viscous liquid of the same texture throughout; *gen*erate, to produce energy or a flow of gas.

**hydr** water or liquids, e.g. de*hydr*ate, to remove water; an*hydr*ous, describes a substance without water.

**hygro** damp or humid, e.g. *hygro*scopic, attracting water from the atmosphere to become damp; *hygro*meter, an instrument that measures the relative humidity of the atmosphere.

**morph** shape or form, e.g. a*morph*ous, describes a substance which is without a crystalline form; poly*morph*ism, existing in different forms.

**photo** light, e.g. *photo*lysis, decomposition caused by light; *photo*halide, any halide which is decomposed by light.

**pneumo** air or gas, e.g. *pneum*atic trough, a trough for the collection of gases.

**pyro** great heat, e.g. *pyro*lysis, decomposition caused by heating; *pyro*meter, a special kind of thermometer for measuring very high temperatures.

**therm** heat, e.g. *thermo*stable, stable when heated; *therm*al, of, or to do with heat; *therm*ometer, an instrument for the quantitative measure of temperature.

# Acknowledgements

There are many standard source books in geology but the author would like to acknowledge his especial indebtedness to the authors of the following books:

BIRKELAND, P.W. and LARSEN, E.E.: *Putnam's geology*, Oxford University Press, New York, 1978

FYFE, W.S.: *Geochemistry*, (Oxford chemistry series), Oxford University Press, Oxford, 1974

GASS, I.G., SMITH, P.J. and WILSON, R.C.L. (editors): *Understanding the Earth*, Open University Press, Milton Keynes, 1971

GUTENBERG, B.: *Internal constitution of the Earth*, Dover Books, New York, 1951

McKERROW, W.S.: *The ecology of fossils: an illustrated guide*, Duckworth, London, 1978

SPENCER, E.W.: *The dynamics of the Earth*, Thomas Crowell, New York, 1972

WELLS, A.F.: *Structural inorganic chemistry*, Oxford University Press, Oxford, 1962 (fourth edition)

WYLLIE, P.J.: *The way the Earth works*, John Wiley, New York, 1976

*The story of the Earth*, Institute of Geological Sciences, London, 1972

# Index

A horizon 23
A 2 horizon 23
áá (lava) 70
abrasion 20
absolute age 121
abundances, cosmic 18
  terrestrial 18
abyssal 34
abyssal hills 34
abyssal plain 34
abyssal zone 36
Acadian 116
acceleration 151
accessory mineral 75
accretionary prism 138
accretion, continental 140
accumulate 151
achondrite 149
acicular 45
acid (rocks) 74
actinolite 57
adaptation (evolutionary) 102
adaptive radiation 102
adularia 56
aeolian weathering 22
aeon 113
aerobic 101
aerolite 149
after-shock 13
age (geological time-division) 113
age, absolute, *see* absolute age 121
  apparent, *see* apparent age 121
agglomerate 69
aggradiation 32
aggrade 32
aggregate 151
Agnatha 109
albite 56
algae 110
alkali 16
alkali feldspars 56
alkali metals 17
alkali pyroxenes 57
alkaline (rocks) 75
allanite 60
allochemical 85
allogenic 81
allothigenic, *see* allothigenous 81
allothigenous 81
allotriomorphic 72
alloy 17
alluvial 24
alluvial fan 24
alluvium 24
almandine 58
alpha quartz 55
Alpine (movements) 116
alpine glacier 28
alter 151
alternate 151
alumina 16
aluminium silicate minerals 59
ammonites, *see* Ammonoidea 107
Ammonoidea 107
ammonoids 107
amorphous 44
Amphibia 109
amphibians, *see* Amphibia 109
amphiboles 57
amphibolite facies 92
amygdale 73
amygdaloidal 73
amygdule, *see* amygdaloidal 73
anaerobic 81, 101
analysis, chemical 17
anatase 49
anatexis 93
ancestral 102
andalusite 59
andesine 56
andesite 76
andradite 58
Angiospermae 110
angiosperms,
  *see* Angiospermae 110
angle, interfacial 40
angle of repose 21
angular 83

angular unconformity,
*see* unconformity 118
anhedral 45
anhydrite 52
anion 15
anisotropic 47
Annelida 105
annelids, *see* Annelida 105
anorthite 57
anorthosite 79
antecedent (drainage) 27
anthophyllite, *see* amphiboles 57
Anthozoa, *see* Coelentera 105
anthracite 89
anticline 124
anticlinorium 124
antiform 124
apatite 49
aphanitic 72
apophysis 66
Appalachian 116
apparent age 121
apparent dip 123
aquifer 146
Arachnida 108
arachnids, *see* Arachnida 108
aragonite 51
arch 38
Archaean 114
Archaeocyatha 104
arenaceous, *see* arenite 85
arenite 85
arête 31
argillaceous 85
argillite 85
arkose 87
arroyo 25
arsenopyrite 50
artesian 146
Arthropoda 108
arthropods, *see* Arthropoda 108
asbestos 61
ash, volcanic 69
asphalt 89
assemblage 100
assimilation 62
asthenosphere 9
astrobleme 131
asymmetrical fold 126
atmosphere 9
atoll 35
atom 152
atomic number 152
atomic weight 152
augen gneiss 97
augite 57
aulacogen 133
aureole, metamorphic 92
authigenic 84
average 152
Aves 109
axial plane (crystallography),
*see* axis 41
axial plane (of fold) 123
axial ratio 41
axis, crystallographic 41
axis of symmetry 42
azurite 51

B horizon 23
back-arc upwelling 138
backshore 37
balance 152
banding, in igneous rocks 63
bar 38
barchan dune 22
barite 52
barrier reef 38
barrier island 38
barytes 52
basalt 77
base 152
base level 24
basement 114
basement complex 114
basic (rocks) 74
basin, structural 125
batholith 64
bathyal 34
bathylith, *see* batholith 64
bauxite 23
beach 37
raised 39
bed 80, 113
bedding 80
convolute 82
cross 82
current 82

bedding, dune 82
  graded 82
bedding-plane 80
beheaded (stream) 25
Belemnoidea 107
belemnites 107
Benioff zone 137
benthonic, *see* benthos 100
benthos 100
Bergschrund 28
berm 37
beryl 60
beta quartz 55
biocoenose 101
biofacies 117
biogenic 101
bioherm 101
biolith 101
bioseries 103
biosphere 98
biostratigraphical 117
biostrome 101
biotite 55
birds, *see* Aves 109
birefringence 47
bitumen 89
bitumenous coal 89
black-band ironstone 88
block (structural) 133
block, volcanic 69
blue john 52
bomb, volcanic 69
borehole 144
boss 65
botryoidal 45
bottomset beds 82
boulder 87
boulder clay 29
boundary 152
box fold 126
Brachiopoda 105
brachiopods, *see* Brachiopoda 105
breccia 87
  volcanic 69
Bryozoa 106
bryozoans, *see* Bryozoa 106
burst, evolutionary 103
butte 33
bytownite 56
C horizon 23
Cainozoic 115
calc tufa 21
calcareous 86
calcareous tufa 21
calc-alkaline (rocks) 75
calcite 51
caldera 68
Caledonian 116
Caledonides 116
caliche 23
Cambrian 114
cancrinite 58
cannel coal 89
canyon 25
  submarine 35
cap rock 144
capture, (river and stream) 25
carbon minerals 48
carbon-14 dating 120
carbonaceous (sediments) 89
carbonate 16
Carboniferous 114
cartilage 152
cassiterite 48
cast 98
cataclasis 94
cataclastic 94
Catastrophism 112
catchment area 146
cation 15
cauldron subsidence 67
cave, sea 37
Cca horizon 23
celestite 52
celestine 52
cell 153
cell, unit 40
cement 84
cementation 84
Cenozoic, *see* Cainozoic 115
central eruption 68
centre of symmetry 42
Cephalopoda 107
cephalopods,
  *see* Cephalopoda 107
chain structure (silicates) 53
chalcedony 55
chalcocite 50

chalcophile 18
chalcopyrite 50
chalk 86
channel 153
charge 153
chemical composition 15
chemical compound 15
chemical equilibrium 17
chemical precipitate,
  *see* precipitation 18
chemical reaction 17
chemical weathering 20
chemistry 153
chert 86
chevron fold 126
chilled margin 66
chilled zone, *see* chilled margin 66
chlorite 61
chondrite 149
chondrule, *see* chondrite 143
Chordata 109
chordates, *see* Chordata 109
chromite 49
chron 113
chronostratigraphical (unit) 113
cinder cone 68
cinnabar 50
cirque 31
cladistics 103
cladogenesis 103
class (taxonomic) 99
classify 153
clast 85
clastic (sediments) 85
clay 88
  red 36
clay-ironstone 88
clay minerals 61
cleavage (rock) 95
  false 95
  flow 95
  fracture 95
  shear 95
  slaty 95
cleavage (mineral) 44
cleavage-plane 44
cleave, *see* cleavage 44
cliff 37
clinometer 148
clinopyroxenes 57
clinozoisite 60
close, *see* closure 123
closed system 19
closure 123
coal 89
  bituminous 89
  cannel 89
  sub-bituminous 89
  rank of 89
coal measures 89
coal seam 89
coarse 153
coast 37
  drowned 39
  emergent 39
  primary 39
  ria 39
  secondary 39
  submerged 39
coastline 37
cobble 87
Coelentera 105
corsite 55
columnar (habit) 45
columnar jointing 67
columnar structure, *see* columnar
  jointing 67
combine 153
community, fossil 100
compaction 84
compass, prismatic 148
competent bed 126
complex 153
component (chemical) 19
compose 153
composite cone 68
compositional zoning 63
composition, chemical 15
compound, chemical 15
concave 153
concentric 153
concentric fold 124
concertina fold 126
concordant (age) 121
concordant (intrusion) 65
concretion 84
conductor 153
conduit (volcanic) 68

cone 153
cone, volcanic 68
  composite 68
  cinder 68
  spatter 69
cone-sheet 67
confining pressure 93
conformable 118
conformity 118
conglomerate 87
connate water 146
consequent (stream) 27
conservative margin 135
consolidation 84
constructive margin 135
contact 148
contact goniometer,
  *see* goniometer 40
contact metamorphism 90
contact zone 92
contemporaneous 118
continent 153
continental accretion 140
continental drift 134
continental rise 34
continental shelf 34
continental slope 34
contour 153
contract 154
convection 154
convection current (in mantle) 142
convergence 103
  zone of 138
convex 154
convolute bedding 82
coral reef 38
corals 105
cordierite 60
cordillera 133
core 9
corrasion 24
correlation 117
correspond to 154
corrie 31
corrosion 20
corundum 49
cosmic abundances,
  *see* abundances of
  elements 18
country-rock 65
crater 150
craton 133
creep, soil 143
crescent marks 83
crest (fold) 124
Cretaceous 115
crevasse 28
Crinoidea 106
crinoids, *see* Crinoidea 106
cristobalite 55
cross bedding 82
cross-cutting 65
cross-section 148
crude 144
crude oil 144
crust 9
Crustacea 108
crustaceans,
  *see* Crustacea 108
cryptocrystalline 44
cryptocrystalline (texture) 72
cryptoexplosion structure 131
crystal 40
  zoned 46
crystal form, *see* form 40
crystal lattice 40
crystal nucleation,
  *see* nucleation of crystals 19
crystal settling 63
crystal system 43
crystal, twin 41
crystalline 40, 44
crystalline texture 72
crystallinity 44
crystallize 40
crystallized 44
crystalloblastic 94
crystallographic axis 41
crystallography 40
crystallographic 40
  axis 41
cube 41
cubic system 43
cuesta 33
cumulate 63
cupola 64
cuprite 48
curie point 14

current, turbidity 36
  density, *see* turbidity current 36
current bedding 82
current ripples,
  *see* ripple mark 83
cwm 31
cycle of erosion 32
cyclic sequence 112
cylindrical fold 126
cyclosilicates 53
cyclothem 112

dacite 76
daughter element 19
decke 130
décollement 130
decompose 154
decussate (texture) 72
dedolomitization 86
deformation 122
deformational fabric,
  *see* tectonite 97
degradation 32
delta 26
deltaic 26
deltaic (environment) 81
dendritic 45
dendrochronology 121
density 154
density current 36
denudation 32
deposit 80
deposition 80
depositional environment 81
derived fossil 111
desert varnish 22
destructive margin 135
detrital 85
devitrification 63
Devonian 114
dextral fault 129
diabase 78
diachronous 117
diagenesis 84
diagram 155
diameter 155
diamond 48
diapir 131
diastem 118
diastrophism 132
Diatomaceae 110
diatoms, *see* Diatomaceae 110
diatreme 69
dichroism 47
differential erosion 21
differential weathering 21
differentiation, magmatic,
  *see* magmatic differentiation 62
diffusion 63
dike, *see* dyke 67
dimension 155
dimorphism 46
dinosaurs 109
diopside 57
diorite 76
dip 123
  true 123
  apparent 123
dip-slip fault 129
dip-slope 33
directed pressure 93
directed stress, *see* stress 122
disconformity 118
discontinuity 10
discordant (age) 121
discordant (intrusion) 65
disharmonic fold 126
displacement (fault) 128
dissolve 155
distillation 155
distributary 26
distribute 155
divergence 102
  zone of 136
diversification 102
diversity (species) 102
division 155
dolerite 78
dolomite 51
dolomitization 86
dome 125
  piercement 131
  salt 131
downthrow 128
downwards 155
downwarp 125
drag fold 126
drainage, antecedent 27

drainage, inconsequent 27
superimposed 27
drainage-pattern 27
dendritic 27
radial 27
rectangular 27
system 25
trellis 27
drift, continental 134
drift (glacial) 31
drill 155
drowned coast 39
drumlin 30
dry valley 25
dune 22
barchan 22
longitudinal 22
parabolic 22
transverse 22
whaleback 22
dune bedding 82
dunite 79
duricrust 23
dust, volcanic 69
dyke 67
neptunean 84
sandstone 84
dyke-swarm 67
dynamic metamorphism 90
dynamothermal metamorphism 90

Earth sciences 9
earth flow 143
Earth, structure of 9
earthquake 12
Echinodermata 106
echinoderms,
*see* Echinodermata 106
Echinoidea 106
echinoids, *see* Echinoidea 106
eclogite 97
eclogite facies 92
effluent (stream) 25
ejectamenta (volcanic) 69
electron 155
element 15
elements, abundances of 18
embayment 39
emergent coast 39
emplacement 64
*en echelon* 155
end-member 46
end-moraine 29
energy 155
engineering geology 138
englacial moraine 29
enrichment 62
enstatite 57
environment, depositional 81
marine 81
sedimentary 81
Eocene 115
eolian weathering 22
eon, *see* aeon 113
epeirogenic movements 125
epibole 117
epicentre 12
epicontinental sea 34
epidotes 60
epifauna 101
epitaxis 46
epizoon 101
epoch 113
equant 45
equigranular (texture) 73
equilibrium 155
equilibrium, chemical 17
equivalent 156
era 113
erode 20
erosion 20
differential 21
erosional cycle 32
erratic, glacial 31
eruption (volcanic) 68
central 68
fissure 68
phreatic 70
escarpment 33
esker 30
essential mineral 75
estuarine (environment) 81
eugeosyncline 132
euhedral 45
Eurypterida 108
eurypterids, *see* Eurypterida 108
eustatic (movements) 36
eutectic point 71

evaporation 156
evaporite 85
evolution 102
  explosive 103
evolutionary burst 103
exfoliation 20
expand 156
explosive evolution 103
exposure 122
exsolution 71
external 156
extinction 102
extrusive 69

face (crystal) 40
face (structural attitude) 127
facies 117
  metamorphic 92
  fossil 111
facies fauna, *see* facies fossil 111
false cleavage 95
family 99
fan, outwash 30
fan fold 124
fault 128
  dextral 129
  dip-slip 129
  listric 130
  normal 128
  oblique-slip 129
  reverse 128
  rotational 129
  sinistral 129
  strike-slip 129
  tear 129
  thrust 129
  transcurrent 129
  wrench 129
fault block 129
fault breccia 129
fault gouge 129
fault plane 128
fault zone 128
fault scarp 129
fauna 98
fayalite 58
feldspars 56
feldspathoids 58
felsic (minerals) 75
felsite 77
felspars, *see* feldspars 56
felspathoids,
  *see* feldspathoids 58
fenster 130
ferromagnesian minerals 55
ferrosilite 57
ferruginous (sediments) 88
fibrous (habit) 45
field work 148
field geology 148
fine 156
fiord 31
fire curtain 70
fire fountain 70
fissile 80
fissure 21
fissure eruption 68
fish, *see* Pisces 109
flexure 133
flint 86
flood-plain 26
flora 110
flow cast 83
flow cleavage 95
flow fold 126
fluorite 52
fluorspar 52
fluvial (environment) 81
fluvio-glacial deposits 30
focus (earthquake) 12
fold 123
  asymmetrical 126
  box 126
  chevron 126
  concentric 124
  concertina 126
  cylindrical 126
  disharmonic 126
  drag 126
  flow 126
  isoclinal 127
  overturned 127
  parallel 124
  recumbent 127
  shear 126
  similar 124
  slip 126
  symmetrical 126

fold, zig-zag 126
fold-axis 123
foliated 45
foliation 95
footwall 128
Foraminifera 104
forams 104
force 156
foreland 133
foreset beds 82
fore-shock 13
foreshore 37
form (crystallographic) 40
formation 113
formula (chemical) 156
forsterite 58
fossil 98
  derived 111
fossilize 98
fossiliferous 98
fossil community 100
fossil record 98
fraction 62
fracture (mineral) 44
fracture (rock) 122
fracture cleavage 95
fragment 156
frequency 156
front 16
frost heave 30
frost wedging 30
fullers' earth 61
fumarole 70

gabbro 76
gal 11
galena 50
gangue 145
garnets 58
gas 156
gas, natural 144
Gastropoda 107
gastropods,
  *see* Gastropoda 107
geanticline 132
gene 156
genotype, *see* type 99
genus 99
geochemical cycle 18
geochemistry 15
geochronology 120
geography 156
geological column 113
geological survey 148
geology 9
geomagnetism 14
geomorphology 32
geophysics 11
geosyncline 132
geotechnics 143
geothermal gradient 93
geothermal heat flow 93
geyser 70
ghost stratigraphy 92
glacial erratic 31
glacial lake 28
glacial period 28
glacial striae 29
glacial striations 29
glaciation 28
glacier 28
  alpine 28
  lake 28
  mountain 28
  piedmont 28
  valley 28
glaciofluvial deposits 30
glaciology 28
glass 77
glassy 72
glauconite 55
glaucophane 57
glide 143
Globigerina ooze 36
Gnathostoma 109
gneiss 97
gold, native 48
Gondwanaland 139
goniometer 40
  contact 40
  reflecting 40
graben 133
gradation 32
grade, metamorphic,
  *see* metamorphic grade 91
graded bedding 82
gradient (stream) 24
grain 72

grain boundary 72
granite 76
granitization 63
granoblastic 94
granodiorite 76
granulite 97
granulite facies 92
graph 157
graphite 48
graptolites,
*see* Graptolithina 109
Graptolithina 109
gravel 87
gravimeter 11
gravitational acceleration 11
gravity 11
specific 44
gravity anomaly 11
gravity meter 11
gravity separation 63
gravity tectonics 130
graywacke, *see* greywacke 87
greensand 87
greenschist facies 92
greywacke 87
grit 87
grossularite 58
groundmass 73
groundwater 146
group (stratigraphical) 113
Gutenberg discontinuity 10
guyot 35
Gymnospermae 110
gymnosperms,
*see* Gymnospermae 110
gypsum 52

habit 45
habitat 100
hade 128
haematite 48
half-life 19
halide 16
halite 52
hand specimen 147
hanging valley 31
hanging wall 128
hardness 44
haüyne 58
heat flow, *see* geothermal heat
flow 93
heave (fault) 128
heave, frost 30
hematite, *see* haematite 48
hemera 117
Hercynian 116
hexagonal system 43
highlands, lunar 150
high quartz, *see* beta quartz 55
hills, abyssal 34
hinge (fold) 123
historical geology 112
Holocene 115
holocrystalline 72
holotype, *see* type 99
homoeomorphy 103
horizon (stratigraphical) 112
horizontal 157
hornblende 57
hornblendite 79
hornfels 96
horst 133
hot spot 142
hyaline 72
hybrid 62
hybridization 62
hydration 157
hydrocarbon minerals 89
hydrogeology 146
hydrology 146
hydrolysis 18
hydrosphere 34
hydrostatic pressure 93
hydrostatic stress, *see* stress 122
hydrothermal 63
hydrothermal deposit 145
hydrous 157
hydroxyl, group 16
Hydrozoa, *see* Coelentera 105
hypabyssal 64
hypidiomorphic 72

Iapetus sea 141
ice age 28
ice-dammed lake 28
ice sheet 28
idioblastic 94
idiomorphic 72

idocrase 60
igneous 62
igneous intrusion 64
igneous (rocks) 62
ignimbrite 70
illite 61
ilmenite 49
imbricate structure 130
impermeable 146
impregnation 84
incised meander 26
inclusion 46
incompetent bed 126
incongruent melting 71
inconsequent (drainage) 27
index fossil 111
index mineral 91
indices (crystallographic) 41
induration 84
influent (stream) 25
injection 64
inlier 118
inorganic 17
inosilicates 54
Insecta 108
insects, *see* Insecta 108
inselberg 33
*in situ* 148
insolation 20
intense 157
intensity (earthquake) 12
interbedded 119
intercalated 119
intercept (crystallographic) 41
interfacial angle 40
interglacial period 28
intergrowth 73
intermediate (rocks) 74
interstratified 119
intraformational 118
intrusion, igneous 64
  multiple 65
intrusive, *see* igneous intrusion 64
inversion, *see* inverted 127
Invertebrata 104
invertebrates, *see* Invertebrata 104
inverted 127
ion 15
ionic radius 15
iridescence 47
iron (meteorite) 149
ironstone 88
island arc 135
island, barrier 38
isobath 35
isochemical process 19
isochron 121
isoclinal fold 127
isograd 91
isomorphous series 46
isopach 122
isopachyte 122
isoseismal 12
isoseismal line 12
isostacy 11
isostatic adjustment 11
isostatic compensation 11
isotherm 93
isotope 19
isotopic age 121
isotropic 47

jet 89
joint 21
joint set 21
joint system 21
Jurassic 115
juvenile water 146

K horizon 23
K–Ar dating 120
kame 30
kaolinite 61
karst topography 33
Kimmerian 116
klippe 130
knick-point 24
kyanite 59

laboratory 147
labradorite 56
laccolith 65
lacustrine (environment) 81
lag 130
lagoon 38
lagoonal (environment) 81
lahar 70
lake, glacial 28

lake, glacier 28
ice-dammed 28
marginal 30
ox-bow 26
lamellae, *see* lamellar 45
lamellar 45
Lamellibranchiata 107
lamellibranchs,
*see* Lamellibranchiata 107
lamina 80
lamination 80
lamprophyre 78
landscape 157
landslide 143
landslip 143
lapilli 69
Laramide 116
lateral moraine 29
laterite 23
laterization 23
lattice, crystal 40
Laurasia 139
lava 70
lava-flow 70
lava tube 70
lava tunnel 70
layer 157
layer lattice silicates 54
layering in igneous rocks 63
lead–lead dating,
*see* uranium–lead dating 120
lead–uranium dating,
*see* uranium–lead dating 120
lepidolite 55
lens (rock) 119
lens (glass) 147
lenticular 119
lepidolite 55
leucite 58
leucocratic 75
levée 26
lignite 89
limb (fold) 123
limestone 86
limnic (environment) 81
limonite 48
lineage 102
lineament 131
lineation 95
liquidus 71
listric fault 130
lithifaction 84
lithification 84
lithofacies 117
lithological, *see* lithology 85
lithology 85
lithophile 18
lithosphere 9
lithostatic pressure 93
lithostratigraphical (unit) 113
*lit-par-lit* 66
littoral 37
load (stream) 24
load cast 83
load pressure 93
lode 145
loess 22
longitudinal dune 22
longitudinal profile 24
lopolith 66
low quartz, *see* alpha quartz 55
lunar 150
lunar highlands 150
lunar regolith 150
lunar soil 150
lustre 47
lustrous, *see* lustre 47
lutaceous, *see* lutite 85
lutite 85
L-waves 12

mafic (minerals) 75
magma 62
magma chamber 64
magmatic differentiation 62
magmatic stoping, *see* stoping 65
magmatism 62
magnesite 51
magnet 157
magnetic anomaly 14
magnetic field 157
magnetic pole 14
magnetism, terrestrial 14
magnetite 49
magnetization, remanent 14
magnetometer 14
magnitude (earthquake) 12
malachite 51

Mammalia 109
mammals 109
mantle 9
mantle rock 23
mantled gneiss dome 131
marble 96
mare (lunar) 150
mare ridge 150
margin 157
  plate 135
  conservative 135
  constructive 135
  destructive 135
marginal lake 30
marginal plateau 34
maria (lunar), *see* mare 150
marine 34, 157
marine environments 81
marine swamp 38
marine terrace 39
marl 88
mascon 150
massive (habit) 45
matrix 73
mature (stream) 24
M-discontinuity,
  *see* Mohorovičić discontinuity 10
meander 26
  incised 26
mechanical weathering 20
medial moraine 29
medium 157
melanocratic 75
melting, incongruent 71
member (stratigraphical) 113
mesa 33
mesosphere 9
Mesozoic 115
meta- (prefix) 90
metal 17
metallogenetic province 145
metamorphic aureole 92
metamorphic facies 92
metamorphic grade 91
metamorphic rocks 90
metamorphic zones 91
metamorphism 90
  contact 90
  dynamic 90
metamorphism, dynamothermal 91
  regional 90
  thermal 90
metaquartzite 96
metasomatism 90
Metazoa 104
meteoric water 146
meteorite 149
  stony 149
method 157
micas 55
micrite 86
microcline 56
microcrystalline (texture) 72
microfauna 99
microfossil 99
microgranite 76
micropalaeontology 99
microplankton 100
microplate 134
microscope 147
  petrological 147
microseism 13
mid-oceanic ridge 35
mid-oceanic ridge basalt,
  *see* MORB 77
migmatite 97
migration (of oil) 144
milligal 11
mine 145
mineral 44
  accessory 75
  essential 75
mineral deposits 145
mineralization 145
mineralogy 44
minor intrusions 67
Miocene 115
miogeosyncline 132
misfit stream 25
Mississippian 114
mobile belt 132
Moho, Mohorovičić discontinuity 10
Mohs' scale 44
molecular structure 15
molecule 15
Mollusca 107
molluscs, *see* Mollusca 107
molybdenite 50

monadnock 33
monocline 124
monoclinic system 43
monomineralic 75
montmorillonite 61
monzonite 78
moraine 29
  englacial 29
  lateral 29
  medial 29
  terminal 29
MORB 77
mould 98
mountain glacier 28
mud cracks 83
mud flow 143
mudstone 88
multiple intrusion 65
muscovite 55
mutation 102
mylonite 97

nappe 130
native element 15
native gold 48
native sulphur 48
natural gas 144
natural selection 102
Nautiloidea 107
nautiloids, *see* Nautiloidea 107
neck, volcanic 68
nekton 100
Neogene 115
nepheline 58
Neptunean dyke 84
neritic 34
neritic zone 34
nesosilicates 53
net slip 128
neutralization 17
NiFe 17
nodule 84
non-depositional unconformity, *see* unconformity 118
non-metal 17
non-sequence 118
normal 40
normal fault 128
nosean 58
notch, wave-cut 37
nucleation of crystals 19
nucleus, atomic 157
nucleus, crystal 19
nuclei 19
nuée ardente 70

oblique-slip fault 129
obsequent (stream) 27
obsidian 77
occur 157
oceanic ridge 35
oceanic trench 35
oceanography 34
octahedron 41
offlap 119
offset 128
off-shore 37
oghurd 22
oil 144
oil shale 144
Oligocene 115
oligoclase 56
olivines 58
ontogeny 102
oolite 86
oolith 86
ooze 36
  *Globigerina* 36
  Radiolarian 36
opencast 145
open system 19
ophiolite 78
ophiolite assemblage 78
ophitic (texture) 73
optical interference 158
optical properties, of minerals 47
order (taxonomic) 99
Ordovician 114
ore 145
ore body 145
organic 17
organic (sediments) 85
organism 98
origin 158
orogen 132
orogenesis, *see* orogeny 132
orogenic belt, *see* orogen 132
orogenic period 116

orogeny 132
orpiment 50
orthite 60
orthochemical 85
orthoclase 56
orthogneiss 97
orthopyroxenes 57
orthoquartzite 87
orthorhombic system 43
orthosilicates, structures,
  *see* nesosilicates 53
oscillate 158
oscillation ripples,
  *see* ripple mark 83
Ostracoda 108
ostracods, *see* Ostracoda 108
outcrop 122
outlier 118
outwash fan 30
outwash plain 31
overburden 145
overlap 119
oversaturated 74
overstep 119
overthrust 129
overturned fold 127
ox-bow lake 26
oxide 16

pahoehoe (lava) 70
palaeobiogeography 100
palaeobotany 110
Palaeocene 115
palaeocurrent 82
palaeoecology 100
Palaeogene 115
palaeogeography 139
palaeomagnetism 14
palaeontology 98
  stratigraphical 111
palaeopole location 14
Palaeozoic 114
paleo-, *see* palaeo-
palimpsest structure 94
paludal (environment) 81
palynology 110
Pangaea 139
panidiomorphic texture 72
parabolic dune 22
paragneiss 97
paralic (environment) 81
parallel 158
parallel fold 124
parameter (crystallographic) 41
parent element 19
particle 158
peat 89
pebble 87
pediment 33
pediplain 32
pedology 23
pegmatite 79
pelagic 100
Pelecypoda 107
pelecypods, *see* Pelecypoda 107
penecontemporaneous 118
peneplain 32
peneplane 32
Pennsylvanian 114
pericline 125
peridotite 79
periglacial 31
period (geological) 113
  glacial 28
periodic table 18
perknite 79
permafrost 30
permeability 146
Permian 114
perthite 56
petrogenesis 62
petrographic province 75
petrography 62
petroleum 144
petrology 62
petrological microscope 147
phacolith 66
phaneritic (texture) 72
phanerocrystalline (texture) 72
Phanerozoic 114
phase 19
phenocryst 73
phlogopite 55
phonolite 77
phosphate 16
photic zone 36
phreatic eruption 70
phyla, *see* phylum 99

phyllite 96
phyllosilicates 54
phylogeny 102
phylum 99
physics 158
phytoplankton 100
piedmont glacier 28
piedmontite 60
piercement dome 131
pillow-lava 70
pinacoid 41
pingo 30
Pisces 109
pisolite 86
pisolith 86
pitch (of fold) 123
pitchblende 48
pitchstone 77
placer deposit 145
plagioclase feldspar 56
plain, abyssal 34
plan 158
plane 158
plane of symmetry 42
plankton 100
plants, *see* palaeobotany 110
plastic 158
plate 134
plate boundary 135
plate margin 135
plate tectonics 134
plateau 32
  marginal 34
  submarine 35
plateau basalt 77
platform (structural) 133
platform, wave-cut 37
Pleistocene 115
pleochroism 47
Pliocene 115
plug, volcanic 68
plume 142
plunge 123
pluton 64
plutonic rocks 64
pneumatolysis 63
poikilitic (texture) 73
poikiloblastic (texture) 94
polar wander 14, 141
polarity, reversed 14
polarized light 158
poles magnetic 14
polymorphism 46
polyphyletic 103
Polyzoa 106
pore-fluid pressure 93
pore space 84
Porifera 104
porosity 84
porous 84
porphyritic (texture) 73
porphyroblast 94
porphyroblastic (texture) 94
porphyry 78
position 158
postkinematic 116
post-orogenic 116
post-tectonic 116
potash feldspar 56
potassium–argon dating 120
Pre-Cambrian 114
precipitation, chemical 18
precipitation (atmospheric) 146
preferred orientation,
  *see* tectonite 97
pressure 93
  confining 93
  directed 93
  hydrostatic 93
  lithostatic 93
  load 93
pore-fluid 93
primary coast 39
principal shock (earthquake) 13
prism 41
  accretionary 138
prismatic (habit) 45
prismatic compass 148
process 158
profile, soil 23
  stream 24
prograde (metamorphism) 91
property 158
proportion 158
Proterozoic 114
proton 159
Protozoa 104
provenance 117

psammite 96
psephite 85
pseudomorph 46
Pteridophyta 110
Pteridospermae 110
ptygmatic (folding) 95
pull-apart zone 136
pumice 69
P-wave 12
pyramid 41
pyramidal (habit) 45
pyriboles 57
pyrite 50
pyroclastic (rocks) 69
pyrolusite 49
pyrometamorphism 90
pyromorphite 49
pyrope 58
pyroxenes 57
pyroxenite 79

quantum evolution 103
quaquaversal 123
quarry 145
quartz 55
quartzite 87, 96
Quaternary 115
quick clay 143

radiation (adaptive) 102
radioactivity 19
Radiolaria 104
Radiolarian ooze 36
radiometric dating 120
raised beach 39
range (of fossil) 111
rank of coal 89
rare earth element 15
ratio 159
Rayleigh wave 12
Rb–Sr dating 120
reaction, chemical 17
reaction rim 46
reaction series 71
reactive 17
realgar 50
Recent 115
recrystallization 93
recumbent fold 127
red beds 88
red clay 36
REE 15
reef 38
  barrier 38
  coral 38
reef (mineral) 145
refer 159
reflecting goniometer 40
reflection seismology 13
refraction seismology 13
refractive index 159
regional metamorphism 90
regolith 23
  lunar 144
regression, marine 119
rejuvenation (stream) 24
relict structure 94
relief 32
remanent magnetization 14
replace 159
replacement deposit 145
repose, angle of 21
Reptilia 109
reptiles, *see* Reptilia 109
reservoir (oil) 144
residual 33
retrograde (metamorphism) 91
reverse fault 128
reversed polarity 14
rheomorphism 93
rhombohedral system 43
rhyolite 76
rhythmic sequence 112
ria 39
ria coast 39
Richter scale, *see* magnitude 12
ridge 159
ridge, mare 144
  oceanic 35
  mid-oceanic 35
riebeckite 57
rift 133
rift valley 133
rigid 159
rill marks 83
rille 150
  normal 150
  sinuous 150

ring-complex 67
ring-dyke 67
ring silicates 53
ripple mark 83
rise, continental 34
river terrace 25
rivers 24–27
roches moutonnées 29
rock 62
rock fall 143
rock glacier 21
rock mechanics 143
rock salt 52
rock-stratigraphical (unit) 113
rotational fault 129
rounded 83
rubidium–strontium dating 120
rudaceous rocks, *see* rudites 85, 87
rudite 85
runoff 146
rutile 49

sabkha 38
salt 17
salt dome 131
sand 87
sandstone 87
sandstone dyke 84
sanidine 56
saturation 74
scheelite 52
schiller 47
schist 97
schistosity 95
schlieren 65
schuppen structure 130
scoria 69
scoriaceous 73
scour-and-fill 82
scree 21
Scyphozoa, *see* Coelentera 105
sea cave 37
sea, epicontinental 34
sea-floor spreading 136
seamount 35
secondary coast 39
section (exposure) 148
section, thin 147
sedentary 101
sediment 80
sedimentary environment 81
sedimentary structures 83
sedimentation 80
sedimentology 80
seif 22
seismic 12
seismograph 12
seismology 12
selenite 52
sequence 112
  cyclic 112
  rhythmic 112
series 159
series (stratigraphical) 113
serpentine 61
sessile 101
shadow zone 12
shale 88
shear cleavage 95
shear fold 126
shear stress, *see* stress 122
sheet silicates 54
shelf 34
shield 133
shield volcano 68
shingle 37
shore 37
shoreline 37
sial 10
siderite (mineral) 51
siderite (meteorite) 149
siderolite 149
siderophile 18
silica 16
silicate 16
silicate structures 53–4
siliceous 87
sill 66
sillimanite 59
silt 88
siltstone 88
Silurian 114
sima 10
similar fold 124
sinistral fault 129
sink-hole 21
site investigation 143
skarn 96

skeleton 159
slate 96
slaty cleavage 95
slickenside 129
slide 130
slip 128
slip fold 126
slope, continental 34
slump 143
smectite 61
smithsonite 51
soda feldspar 56
sodalite 58
sodium feldspar, *see* feldspar 56
soil 23
soil creep 143
soil horizon 23
soil mechanics,
  *see* rock mechanics 143
soil profile 23
soil, lunar 150
sole marks 83
solid solution 46
solid solution series 46
solidus 71
solifluction 30
solution 159
sorosilicates 53
sorting 80
sparite 86
spatter cone 69
speciation 103
species 99
specific gravity 44
specimen 147
Spermatophyta 110
spermatophytes,
  *see* Spermatophyta 110
spessartite 58
sphalerite 50
sphene 59
spherulite 73
spilite 78
spinels 49
spit 38
sponges, *see* Porifera 104
spreading, sea-floor 136
spring 146
spur, truncated 31
stable 159
stack (sea) 38
stage (stratigraphical) 113
stalactite 21
stalagmite 21
states of matter 159
staurolite 59
stock 65
stockwork 145
stony (meteorite) 149
stony-iron (meteorite) 149
stoping 65
strain 122
strata, *see* stratum 80
stratification 80
stratified 80
stratigraphical palaeontology 111
stratigraphy 112
stratovolcano 68
stratum 80
streak 44
stream, beheaded 25
  capture 25
  consequent 27
  effluent 25
  influent 25
  misfit 25
  obsequent 27
  subsequent 27
stream gradient 24
stream profile 24
streams 24–7
stress 122
  directed 122
  hydrostatic 122
striae, glacial 29
striations, glacial 29
strike 123
strike-slip fault 129
stromatolites 101
structure (geological) 122
structure, molecular 15
structures, sedimentary 83
subangular 83
sub-bituminous coal 89
subduction zone 137
sub-era 113
subfamily 99
subhedral 45

subjacent 65
submarine 35
submarine canyon 35
submarine plateau 35
submerged coast 39
subrounded 83
subsequent (stream) 27
subsidence 125
subsoil 23
subsurface 159
succession 112
sulphate 16
sulphide 16
sulphur, native 48
supercooling 19
superfamily 99
superimposed (drainage) 27
superposition (principle of) 112
survey 148
  geological 148
suspension load 24
suture 140
swallow-hole 21
swamp, marine 38
swash marks 83
S-wave 12
syenite 76
syenodiorite 78
sylvite 52
symmetrical fold 126
symmetry 159
  (crystal) 42
  axis of 42
  centre of 42
  plane of 42
symmetry element 42
symplectic (texture) 73
syncline 124
synclinorium 124
synform 124
synkinematic 116
synorogenic 116
syntectonic 116
syntexis 93
system open 19
  closed 19
  geological 113

tabular (habit) 45
Taconic 116
talc 61
talus 21
taphonomy 99
taphrogenesis 133
taxon 99
taxonomy 99
tear fault 129
tectonics 122
  plate 134
tectonite 97
tectosilicates 54
tektite 149
temperature 160
temperature–composition
  diagram 71
tensile stress, *see* stress 122
tension gash 21
terminal moraine 29
term 160
ternary diagram 71
terra rossa 33
terrae (lunar) 150
terrace, marine 39
  river 25
terrestrial 81
terrestrial abundances (of
  elements), *see* abundances of
  elements 18
terrestrial magnetism 14
terrigenous (sediments) 85
Tertiary 115
Tethys 141
tetragonal system 43
tetrahedron 41
texture (rock) 72
Thalweg 24
theory 160
thermal metamorphism 90
thin section 147
tholeiite 77
thorium–lead dating,
  *see* uranium–lead dating 120
throw 128
thrust 129
thrust fault 129
thrust plane 129
thrust sheet 130
tide 37

tide mark 37
till 29
tillite 29
time, geological 120
time plane 112
tissue 160
titanite 59
tongue 66
topaz 59
topset beds 82
tourmaline 60
trace element 15
trace fossil 98
trachyte 77
traction load 24
transcurrent fault 129
transform fault 136
transgression, marine 119
transient (evolution) 103
transport 21
transportation 21
transverse dune 22
trap, stratigraphical 144
structural 144
tree-ring dating 121
tremolite 57
tremor 13
trench, ocean 35
trend (evolutionary) 103
triangular diagram 71
Trias 115
Triassic 115
triclinic system 43
tridymite 55
trigonal system 43
Trilobita 108
trilobites, *see* Trilobita 108
triple junction 135
trough (fold) 124
trough faulting 133
true dip, *see* dip 123
truncated spur 31
tsunami 36
tufa, *see* calcareous tufa 21
tuff 69
turbidite, *see* turbidity current 36
turbidity current 36
twin crystal 41
type-area 118
type locality 118
type (specimen) 99

ultrabasic (rocks) 75
ultramafic (rocks) 75
unconformity 118
unconsolidated 84
undersaturated 74
Uniformitarianism 112
uninverted 127
unit 160
unit cell 40
unit form 41
unmetamorphosed 90
unreactive, *see* reactive 17
unstratified 80
unweathered 20
U–Pb dating, *see* uranium–lead dating 120
uplift 125
upthrow 128
upward 160
upwarp 125
uraninite 48
uranium–lead dating 120
U-shaped valley 31
uvarovite 58

valley glacier 28
valley, dry 25
hanging 31
U-shaped 31
variation 160
Variscan 116
varnish, desert 22
varve-count 121
vary 160
vascular plants 110
vein 145
velocity 160
vent, volcanic 68
ventifact 22
Vermes 105
vermiculite 61
Vertebrata 109
vertebrates, *see* Vertebrata 109
vertical 160
vesicle 73
vesicular 73

vesuvianite 60
vibrate 160
volatile (constituent) 18
volcanic, *see* volcano 68
volcanic agglomerate 69
volcanic ash 69
volcanic block 69
volcanic bomb 69
volcanic breccia 69
volcanic cone 68
volcanic conduit 68
volcanic dust 69
volcanic eruption 68
volcanic neck 68
volcanic plug 68
volcanic vent 68
volcanism, *see* volcano 68
volcano 68
  shield 68
volume 161
V-shaped valley 24
vulcanism, *see* volcano 68

wadi 25
wall rock 145
warping 125
washout 82
waterfall 25
water, connate 146
  juvenile 146
  meteoric 146
watershed 25
water table 146
wave base 36
wave-cut platform 37
wave-cut notch 37
wave motion 161
wavelength 161
way-up 127
weathering 20
  chemical 20
  mechanical 20
  aeolian 22
  differential 21
  eolian, *see* aeolian 22
Weichert-Gutenberg
  discontinuity 10
well-logging 144
well-rounded 83
whaleback dune 22
white mica, *see* muscovite 55
window 130
witherite 51
wolframite 52
worms 105, *see also* Annelida 105
wrench fault 129
wrinkle ridge 150

xenoblastic 94
xenocryst 73
xenolith 65

young (*v*) 127

zeolites 61
zeugen 21
zig-zag fold 126
zinc blende 50
zircon 59
zoisite 60
zone 161
zone (crystallographic) 40
  (metamorphic) 91
  neritic 34
  (stratigraphical) 117
zone fossil 111
zoned crystal 46
zone of convergence 138
zone of divergence 136
zoning, compositional 63

# Divisions of geological time*

| era and sub-era | | period | epoch | | age/stage | ages (Ma) |
|---|---|---|---|---|---|---|
| CENOZOIC | Quaternary | ——— | Holocene | | Flandrian/Versilian | |
| | | | Pleistocene | Upper | Tyrrhenian<br>Milazzian<br>Sicilian | |
| | | | | Lower | Emilian<br>Calabrian | 2 |
| | Tertiary | Neogene | Pliocene | Upper | Piacenzian | |
| | | | | Lower | Zanclian | |
| | | | Miocene | Upper | Messinian<br>Tortonian | |
| | | | | Middle | Seravillian<br>Langhian | |
| | | | | Lower | Burdigalian<br>Aquitanian | |
| | | Palaeogene | Oligocene | Upper | Chattian | |
| | | | | Lower | Rupelian<br>'Lattorfian' | |
| | | | Eocene | Upper | Priabonian<br>Bartonian | |
| | | | | Middle | Lutetian | |
| | | | | Lower | Ypresian | |
| | | | Palaeocene | Upper | Thanetian | |
| | | | | Lower | Danian | 65 |
| MESOZOIC | | Cretaceous | | Upper | Maastrichtian<br>Campanian<br>Santonian<br>Coniacian<br>Turonian<br>Cenomanian | |
| | | | | Lower | Albian<br>Aptian<br>Barremian<br>Hauterivian<br>Valanginian<br>Berriasian | 140 |
| | | Jurassic | | Upper | Portlandian/Tithonian/Volgian<br>Kimmeridgian<br>Oxfordian | |
| | | | | Middle | Callovian<br>Bathonian<br>Bajocian | |
| | | | | Lower | Toarcian<br>Pliensbachian<br>Sinemurian<br>Hettangian | 195 |
| | | Triassic | | Upper | Rhaetian<br>Norian<br>Carnian | |
| | | | | Middle | Ladinian<br>Anisian | |
| | | | | Lower | Scythian | 230 |

*In American charts of geological time, the Quaternary and Tertiary are known as periods rather than sub-eras. Also, the Carboniferous period of the Paleozoic era is divided into two epochs, the Pennsylvanian and Mississippian.

Watt, Alec

Barnes & noble thesau
thesaurus of geology